Artificial Intelligence and Machine Learning in Healthcare

Ankur Saxena • Shivani Chandra

Artificial Intelligence and Machine Learning in Healthcare

Springer

Ankur Saxena
Amity University
Noida, Uttar Pradesh, India

Shivani Chandra
Amity University
Noida, Uttar Pradesh, India

ISBN 978-981-16-0810-0 ISBN 978-981-16-0811-7 (eBook)
https://doi.org/10.1007/978-981-16-0811-7

This Springer imprint is published by the registered company Springer Nature Singapore Pte Ltd.
The registered company address is: 152 Beach Road, #21-01/04 Gateway East, Singapore 189721, Singapore

Contents

About the Authors

Dr. Ankur Saxena is currently working as an Assistant Professor at Amity University, Noida, Uttar Pradesh. He has been teaching graduate and post-graduate students for more than 15 years and has 3 years of industrial experience in software development. He has published 5 books and more than 40 research articles in reputed journals and is an editorial board member and reviewer for several journals. His research interests include cloud computing, big data, machine learning, evolutionary algorithms, software frameworks, design and analysis of algorithms, and biometric identification.

Dr. Shivani Chandra is an Assistant Professor at Amity Institute of Biotechnology, Amity University, Uttar Pradesh, Noida. She has more than 20 years of experience in biotechnology and molecular biology. Her research interests include genomics analysis, computational biology, and bioinformatics data analysis. She has submitted more than 4000 clones to the NCBI GenBank and was one of the key players in the Rice Genome Sequencing Project. She has published several research articles in genome sequencing, comparative genomics, and genome analysis in reputed journals. She has more than 15 years of teaching experience in computational biology, molecular biology, genetics, recombinant DNA technology, and bioinformatics.

List of Figures

List of Tables

Practical Applications of Artificial Intelligence for Disease Prognosis and Management

1

Abstract

Artificial intelligence (AI) is an emerging field, which provides enhanced capabilities of decision-making to the machines. The extremely popular application of machine learning approaches in the area of disease prognosis and management is the "precision medicine," which can be described as deciding the best treatment options based on features, such as attributes of the patients and the treatment undertaken. By knowing the hidden pattern of the data and its knowledge, computers can predict the future events. Thus, it helps the machine to learn effortlessly without any human intervention and makes easy to do complicated decision- making process. The objective of this chapter is to comprehend and explore the applications of artificial intelligence for the better management of the early prognosis and treatment protocols for diseases. The focus of the chapter will be towards the application of artificial intelligence techniques to medical data management. These techniques can analyse different types of data retrieved from patient samples, such as structured images, features based on patient vitals for predicting the probability of the outcome of a disease and design a better treatment protocol.

Keywords

Artificial intelligence · Disease management · Disease prognosis · MATLAB · Predictive modelling

1.1 Overview of Application of AI in Disease Management

Artificial intelligence (AI) is an intelligence technology that is artificially programmed by humans to mimic like human. This artificial intelligence gets integrated with computer system that is called AI system, which ultimately functions

A. Saxena, S. Chandra, *Artificial Intelligence and Machine Learning in Healthcare*, https://doi.org/10.1007/978-981-16-0811-7_1

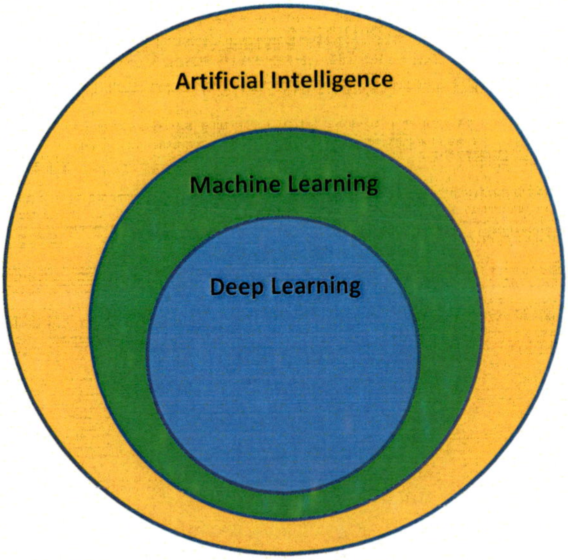

Fig. 1.1 Artificial intelligence, machine learning, and deep learning

as the "thinking machine" (Wu 2019). This AI system responds to the stimulation consistent with the traditional responses made by human and is provided with the human capacity for contemplation, intention, and judgment. The AI helps the computer/system to make decision, which normally requires human-level expertise and helps people to predict or forecast the outcome or deal with issues as they come up (Vijay 2013). According to the Darrell M. West (West 2018) report, the AI system have three qualities, i.e. intelligence, intentionality and adaptability. An intelligent system can be build using the umbrella of techniques under artificial intelligence (AI) that performs human activities well (Fig. 1.1).

The machine learning (ML) is the subset of an artificial intelligence that helps a computer/system to learn from the environment automatically without any human intervention and applies that learning to make better decisions. Machine learning uses its various algorithms or techniques to learn, characterize and improve the data, so that it predicts better outcomes. The ML techniques/algorithms find the patterns first and then perform the action based on these patterns. Machine learning can be classified under four categories: (a) supervised learning, (b) unsupervised learning, (c) semi-supervised learning and d) reinforcement learning.

Supervised Learning

Supervised learning can be defined to be a type of machine learning, where both the input and the output is provided to the system (Akella 2020). The algorithm works by training the labelled data in a manner that the machine is able to learn and develop patterns between the input and the output data. It finds the pattern that tell us how we can categorize or classify datapoints in data. The labelled data means known description, which is given to instances of data. For example, there are 20 different people who have different symptoms with cancer test results. According to the test results, we can place a tag or label to each patient, whether he/she is cancer positive or negative. Hence, the labelled data provides a shape to output. So, the process of supervised learning signifies that the machine will learn the pattern and classify the data. Same patterns can be used to find the unseen data. Supervised learning can be split into two forms:

(a) *Classification*: It is the supervised machine learning algorithm that classify the input data from pre-defined classes. The algorithms help to predict the categorial output from the labelled data.
(b) *Regression*: It is the supervised learning algorithm that finds the relationship between the variables/features. The regression algorithm predicts the output when the input is given by finding the relationship b/w the features.

The list of supervised learning algorithms/techniques:

- K-nearest neighbour.
- Support vector machine.
- Naive Bayes.
- Decision tree.
- Random forest.
- Linear regression.
- Logistic regression.
- Linear discriminant analysis.

Unsupervised Learning

The second category of the algorithms is the unsupervised learning. In this case, the machine is only provided with the input values/data, and there is no fixed output provided to the system. This is the reason why the unsupervised learning algorithms doesn't have labelled data. It predicts the output by finding the hidden pattern from the input data. In comparison with the supervised learning, the problem is not properly defined in unsupervised learning. It is also called lazy learning, but it can find a new way to solve the problem and predict the output from its own. In the process of unsupervised learning of machine, an unlabelled input data is provided. This data is used by unsupervised learning algorithm to hypothesize a pattern within the data on its own. Using the pattern, datapoint instances are grouped. Data matching with the similar pattern and group is predicted as the output (Akella 2020). It is applicable in anomaly detection, segmentation, etc. The unsupervised learning is of two types:

(a) *Clustering*: This method of unsupervised learning relies on making clusters from the input data. The datapoints that have similarities will make clusters, and using those clusters, we will be able to make predictions.

(b) *Association*: The second method of unsupervised learning is association, in which the algorithms find the rules from the input data and make prediction from the data.

The list of algorithms of unsupervised learning is as follows:

- Hierarchical clustering.
- K-means clustering.
- Principal component analysis.
- Neural networks.
- DBSCAN.

Semi-Supervised Learning

The third category of algorithm for machine learning is the semi-supervised learning, which uses both labelled and unlabelled data to build a prediction model. The difference between the above algorithms and semi-supervised method is that the unlabelled data is more in number to the labelled data. Semi-supervised method thus can be considered a fusion of both the supervised and unsupervised learning. The algorithms of the semi-supervised learning include the following: heuristic approaches, generative models, cluster assumption, graph-based methods, lower-density separation, manifold assumption and continuity assumption.

Reinforcement Learning

The final category of machine learning algorithms is the reinforcement learning that focuses on finding the best way to take in a situation that will maximize correct outcome in a situation. The decisions are made sequence-wise. In each step that the algorithm takes on the path to total outcomes, it can either have a positive or negative output. The overall result is thus the sum of all positive and negative outcomes along the path. The algorithm goal is to find the best way that maximizes the outcome. The algorithms that come under reinforcement learning include Q-learning, policy iteration and deep Q network.

Deep learning (DL) is considered to be a subclass of machine learning algorithms; it can also be called the higher version of machine learning that forms multilayer progressively to excerpt features/attributes from the input data to make better and more reliable predictions. The deep learning (DL) provides the computer/system the ability to understand the data from a lower level all the way up to the chain and helps to improve the performance over time and make decisions at any time (Wu 2019). Deep learning (DL) methods are able to work on both supervised and unsupervised tasks. It makes resemblance to many brain development theories of the human brain. The deep learning (DL) algorithms/techniques include:

- Artificial neural network.
- Convolutional neural network.
- Multiple linear regression.
- Gradient descent.

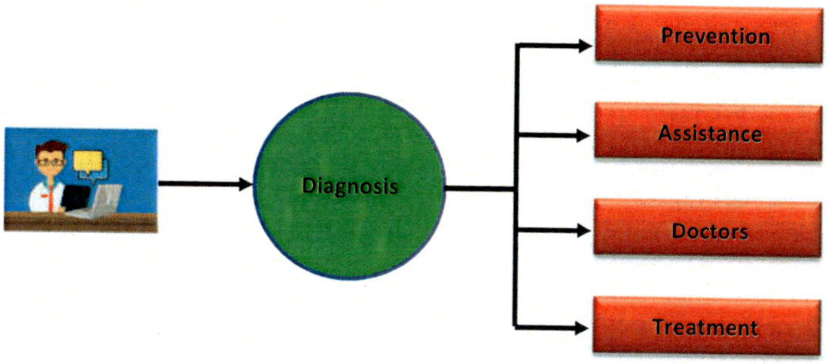

Fig. 1.2 AI in disease management

Machine learning (ML) models in some cases still need intervention by humans to get favourable outcome. The deep learning (DL) models uses artificial neural network (ANN); it is designed in a resemblance of biological neural network of the human brain. It analyses the structure logically like the human draws inference.

AI, along with ML and DL methods, has led to a huge impact in the healthcare sector. These approaches have allowed to undertake a number of innovations in the domains of disease management, i.e. identification of biomarker of a disease, disease prognosis and diagnosis, drug development and personalized medicine (Fig. 1.2).

The rapid growing availability of healthcare medical data and advancement in technologies, such as big data, has led to achieving the applications of AI in the disease management (Datta et al. 2019). Artificial intelligence (AI) in disease management helps people to have healthier and longer lives. Machine learning (ML) techniques/algorithms, such as those discussed in the above section, have the capability of solving complex healthcare concerns, by separating hidden healthcare information from an enormous dearth of data quantity of data and help in making informed decisions that help in disease management. An important application of machine learning (ML) is the process of structuring the different types of medical data, such as genomic data, imaging data, etc., and accurately investigating the same. AI/ML uses these types of data and the extract the potential features/attributes that can be assisted for disease screening/diagnosis and prediction purposes and decision-making in disease treatment and management in real time. This makes the work of clinicians less complex and provides better understanding that helps in managing and treating the disease severity in the patients.

Another important application of AI/ML has been observed in various stages of drug development process. The utility of AI/ML techniques lies primarily in the identification of drug targets and their validation, the repurposing of the drugs, the design of new drugs, improvisation of the R&D processes of drug development and analysing biomedicine data. AI/ML can lead to better decision-making for all these applications, which would lead to faster clinical trials and less expensive drugs.

Another field, wherein AI/ML plays a major role, is in the identification of biomarkers for the disease and personalized medicine. The output generated from the methods such as supervised ML can be successfully employed in the diagnosis of disease under the classification of subgroups, efficacy of drugs and ADMET prediction (Mak and Pichika 2019). The unsupervised ML methods can be used in the discovery of disease subtype, using clustering, and discovery of targets, using feature selection methods. The results from reinforcement ML can be used for de novo drug design and experimental designs, through modelling and quantum chemistry (Mak and Pichika 2019). The AL/ML algorithms and the methodology are immensely useful in providing with the ability of discovering better compounds, which could further be developed in newer drugs or drug repurposing, leading to a cheap and effective drug development process. Accordingly, the response of the patient to the drugs can be monitored effectively, and treatment can be planned individual specific. AI/ML approaches can help us automate the patient response to a line of treatment and manage the disease effectively. This is achieved by the AI system, through a process of learning from the previous records and patient profiles. After the learning process is completed, the algorithm then performs a comparison, by analysing patters, and thus generates a better protocol and treatment plan. This can be further strengthened by the identification of biomarkers through AI/ML. The biomarker identification aids in an absolute clarity for better understanding of the disease prognosis, as the presence of the biomarkers indicates the occurrence of the disease and makes it easy for the clinicians to decide on the appropriate treatment. This makes the process of diagnosing a disease *fast and easy, but discovering the biomarker of disease is still very hard and it is also a very expensive process at the same point.* AI plays an important role in the automation of the various processes in disease management, which is discussed in the sections below.

1.1.1 Disease Prognosis and Diagnosis

Disease prognosis and diagnosis are one of the most important aspects of disease management. However, with the traditional clinical practices, predicting an outcome of an underlying condition is very difficult (Croft et al. 2015). To solve this problem, the disease prognosis was coming into trend that helps to predict the likelihood of future outcomes of the onset of disease that was more useful for clinicians to give the proper treatment to patients. The disease prognosis was still leading the limitation in generalizability to local settings and validity of the study (Lee et al. 2017).

AI/ML leads to a robust transformation in the medical practice. It is helping the doctors and clinicians to diagnose patients more accurately, making predictions and prognosis about the patient's future health, and help to suggest the required treatment of a disease. Artificial intelligence (AI) may create many fears, mainly in the clinical setting, that AI could lead to the reduction of clinician expertise. However, there has been a better acceptance and scope of AI/ML in the clinical setting, where it is believed that these approaches will in turn benefit the clinicians to make better and informed decisions. In a scenario, where a patient is suffering from multiple comorbidities, relating the diagnosis to both the physical and genetic features could be hard and time-consuming. In such cases, AI/ML could aid with the clinicians quantitatively and qualitatively for an early detection and treatment plan

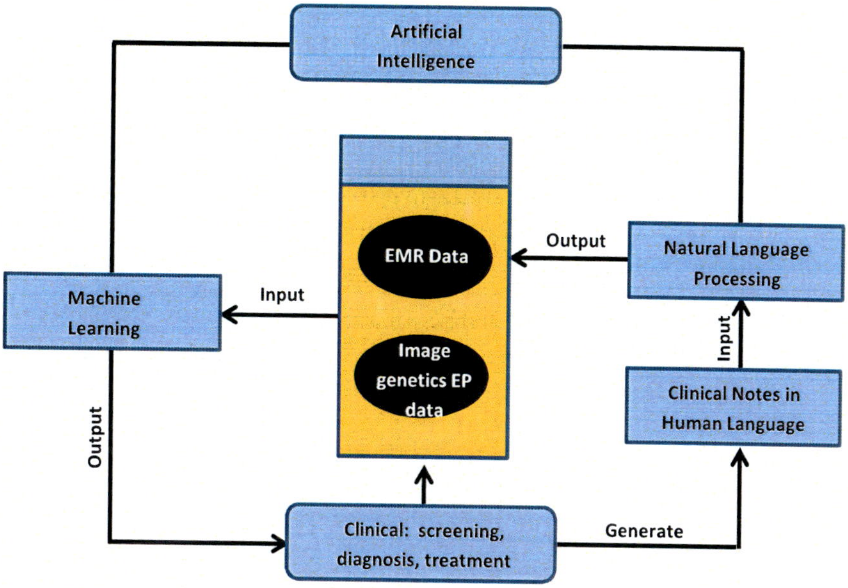

Fig. 1.3 Overall process of the application of AI in disease prognosis and diagnosis

for a better outcome. Machine learning (ML) algorithms can be useful in the principle for analysing the clinical data, such the data from the electronic health records (EHR), imaging data and genetic data. The main objective of these techniques is to cluster the patient's characteristics for predicting the disease occurrence and outcome. The alternative approach is to analyse the information from the unstructured data generated from the clinical notes or medical journals. This is achieved through the natural language processing (NLP) methods. This approach is useful for converting the raw unstructured information into a machine-readable format for analysis using sophisticated AI/ML algorithms. Application of these methods, AI/ML can create a system, which is more accurate and efficient for making the diagnosis and treatment protocols. For example: AI is used to obtain phenotypic characteristics from case reports to enhance the accuracy of diagnosis for congenital abnormalities (Fig. 1.3).

In the recent decade, there have been much advances and better treatment modalities for the major life-threatening diseases, such as cardiovascular disorders, neurological disorders and cancers (Zheng et al. 2005). There have been reports where AI is able to make an early diagnosis of cardiovascular disorders using the image data of cardiac patients (Dilsizian and Siegel 2014), such as CT scan and ECG scan data. Likewise, AI/ML has a tremendous potential in the management of stroke-related cases, through early prediction, forecasting and prognosis of stroke, for better treatment and assessment. A device was built that helps in the early diagnosis of stroke, using machine learning algorithms (PCA and fuzzy) that learn and understand the patients in human detection phase, and starting stroke phase, the device/

model was able to detect the stroke and can stimulate and assess the medical action, thus making it feasible (Villar et al. 2015). The confirmatory accurate diagnosis and also treatment in neurological disorders are still lacking. Here, artificial intelligence (AI) in neurosciences provides the better understanding of intelligent working of biological brains. AI/ML aims to mimic the human thinking functions. A study was done to predict the neurological disorders in the people and its conditions using machine learning techniques (KNN, HMM, MLP and Bayes), the predictions made on brain oscillation characteristics, sleeping and neonatal data. Many AI/ML-based automated CAD systems have been built that include various classifier algorithms (SVM, ANN, KNN, etc.) developed for different neurological disorders, such as Parkinson's disease, Alzheimer's disease, etc. (Raghavendra et al. 2019).

AI/ML has been utilized to develop devices that are useful in monitoring tremors which are helpful for better detection of epileptic conditions. The application of AI integrated electroencephalogram learning can help in preventing the sudden unexpected death in epilepsy (SUDEP) (Patel et al. 2019). A study says artificial neural network (ANN) gave the highest accuracy (>95) in Parkinson's disease detection, and the SVM algorithm knows to be a successful algorithm in predicting the severity of symptoms (Belić et al. 2019).

In oncology, the IBM Watson has developed a system that can be reliable for the identification of cancer at an early stage. Different algorithms and classifiers have the potential of offering prognosis for cancer patients (Huang et al. 2020). For example, this clinical image can be examined using AI/ML techniques for recognizing the skin cancer subtypes. The quick diagnoses and prognosis can potentially reach throughout the recovering of the analysis measures on electronic health records (EHR) or electrophysical (EP), imaging and genetic data, and this shows the power of AI. Apart from these three main diseases, the AI/ML techniques had been used in other diseases also: for example, AI/ML is able to examine the ocular image data for the diagnosis of all cataract diseases.

1.1.2 AI in Identification of Biomarker of Disease

Biomarkers are defined as the quantifiable entities that are observed in biological fluids that provide an understanding of whether a patient has a disease. The biomarkers are the measurable indicators that help to give an idea about the presence or severity of a disease, infection or exposure. Thus, biomarker is very useful for disease diagnosis/prognosis, drug design and development precision medicine. The biomarkers play various roles for curing a disease of patients by knowing the exact stage of disease (Reddy 2019). It can be classified as:

- Prognostic biomarker.
- Diagnostic biomarker.
- Risk biomarker.
- Predictive biomarker.

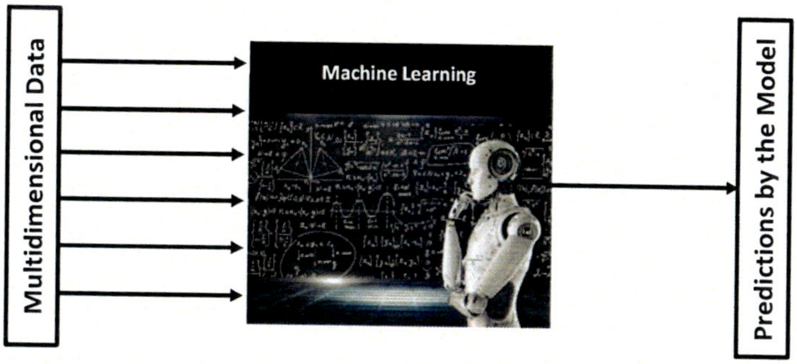

Fig. 1.4 AI/ML techniques help to identify the biomarker of a disease from multidimensional data

But the identification and validation of biomarkers is very time-consuming and expensive. The identification of biomarkers is one of the most important steps for studying disease severity and involves the screening of a number of molecules that could be potentially be considered as biomarkers (Schmitt 2020). Here, artificial intelligence (AI) can automate the process of identifying the suitable candidates and helps clinicians/doctors to know the statistical difference between diseased and healthy humans. As there is large amount of medical data available on biomarkers that help ML/AI techniques to collect this vast amount of data and can make inferences to get the potential candidate as the biomarkers of disease.

The machine learning (ML) has various applications using liquid biopsy data such as disease diagnosis, prognosis and prediction, and now liquid biopsy approach helps to identify a vast number of biomarkers from bodily fluids, such as blood, saliva, urine, tears, faeces and sweat. Using this approach, various sensors have been built with having sufficient sensitivity and specificity to identify novel biomarkers for clinical samples (Ko et al. 2019). By using computational tools, it can decode the biomarker of patient disease and helps in patient treatment. This task however is highly challenging, since there exists a higher variability of the expression of biomarkers in different individuals. This is due to the fact that many disorders are heterogenous in nature and can exhibit multiple biomarkers several times at a point, in a study machine learning techniques said to be helpful in identify the potential biomarker of a particular disease from these multiplexed/multidimensional data, the ML techniques like SVM, decision trees, and random forests, that perform better in terms of specificity and sensitivity of biomarkers in many applications (Ko et al. 2019) (Fig. 1.4).

With the help of liquid biopsy data and AI, the biomarker discovery has been improved a lot. Nowadays, the data-driven biomarker discovery, using various AI/ML methods, has been trending. The various feature/data extraction techniques used pattern matching and speech identification for unstructured data across the public databases, such as KEGG, gene ontology, etc. (George 2020). Using ML techniques such as k-means and hierarchical clustering analyses on lung cancer and ovarian cancer, GEO datasets are able to classify the potential genes from the pool of genes, in a study network that was built to identify the most potential biomarker CREB1 that helps to know the progression of prostate cancer (Pawar et al. 2020).

Biomarkers are also used to predict the longevity of a person, called longevity biomarker. The longevity biomarker combined with AL/ML techniques has the ability to cure the age-associated disorders and helps to improve the lifespan (Colangelo 2020). In a recent scenario, digital biomarker is in fashion, such as Fitbit, Misfit, Jawbone, Apple Health, Sleep as Android, WIWE, Moca Care and Skeeper—in other words, fitness trackers, step counters, health apps, sleep sensors, pocket ECG, blood pressure or other health parameter measuring devices are very popular nowadays (The Medical Futurist 2018). The digital biomarkers are the data that the consumers instantly get the information about the health and disease management from digital health technology that describes, controls and predicts the health-related outcome. The AI/ML technologies collect and analyse and make patterns from a large amount of data (e.g. EHR records) to create a digital biomarker (McCarthy 2020).

1.1.3 AI in Drug Development

Drug development is a process, which starts by generating information from high-throughput screening of compounds and fragments through computational modelling protocols. The process starts with the identification of the drug targets or novel compounds, showing relevant biological activity. These compounds or "hits" are obtained through high-throughput screening of several resources and libraries of chemical compounds (Mak and Pichika 2019). Further, some of these compounds can be also be obtained from natural products from plant/bacterial or fungal sources (Zhu et al. 2013). The process continues by screening these hits in cell assays that depict the disease state in the model organisms, which can depict the efficacy and usability of the compound. This process is known as the target validation. The next step is to identify the lead compounds for the drug development process (Anderson 2011). The drug development process is a multistep protocol, which is laid down by stringent guidelines, and a lot of hurdles are faced by the manufacturers for the improvement of the efficiency of R&D (Mak and Pichika 2019). The increased R&D cost and higher attrition rate in developing the new drugs during drug development process were occurred as a big challenge for pharmaceutical companies. The major part of attrition was occurred in the preclinical development stages that include clinical safety and efficacy that are followed by studies on toxicity, bioavailability, and emphasis on the pharmacokinetics. The drug development process is leading an expensive process, due to the increasing size of clinical trials that are followed according to the FDA rules (Alanine et al. 2003).

Artificial intelligence (AI) is emerging as a versatile tool, leading an era of a cheaper, faster and more effective approach in drug development. The AI/ML techniques, after integrating with pharmaceutical companies, help a lot in drug development process. It is applicable in every stages of drug development process, which had improved and made faster the drug development process with low-cost

Fig. 1.5 AI in drug development

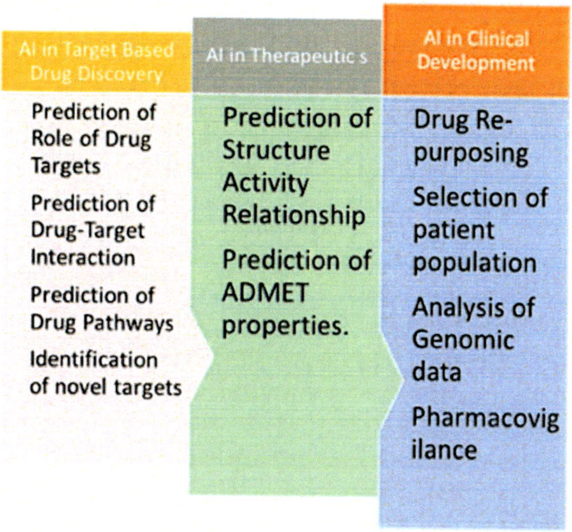

time. AI/ML techniques help to identify and validate the drug targets, de novo drug design and drug repurposing more accurately. Using artificial intelligence (AI), R&D efficiency has been also improved. This can be achieved by the collection and analysis of the biomedical data, which can help in better decision-making clinical trials. The potential uses of AI offer the chances of solving the inadequacies and ambiguities that are witnessed by the traditional drug discovery protocols and avoid any bias generated due to the human intervention. The role of AI in drug development can be further detailed through Fig. 1.5.

The application of AI in the field of drug development is observed in the prediction of synthetic paths of drug-like compounds (Merk et al. 2018), identification of the pharmacological properties, characterization and efficacy testing of protein receptors and analysing the association between the drugs and the targets (Schneider 2017). Using AI/ML techniques, it is possible to identify pathways of the targets, using the omics-based data, and this could lead to generating new biomarkers and identify the therapeutic targets. This will further pave the way for personalized medicine and uncover better relationship between the disease and the drug efficacy. Deep learning methods had shown excellent response in suggesting prospective drug compounds and precisely predict the drug properties, by analysing the toxicity of the drugs for risks in its administration. AI has been pivotal in solving number of problems of analysing the larger datasets and can help improvise the screening of number of compounds, which is a lot time-consuming process (Mohs and Greig 2017). Some of the examples of AI in drug development can be seen through a study, where the therapeutic targets were predicted using computational approaches, which were referred to as open targets. This is a large collection of the disease and gene association. Further, in this study, it was predicted that using a neural network classifier of more than 71% has the maximum potential of better animal models (Ferrero et al. 2017). Another major milestone was achieved by the IBM Watson for the Drug Discovery Group. This group has developed an AI-based platform, which was able to identify RNA-binding proteins, which were linked to the occurrence of amyotrophic lateral sclerosis (ALS) (Bakkar et al. 2018).

Through previous literature, thus, we can infer that AI/ML plays a major role in the process of drug development at many levels. The application of AI/ML can lead to a much faster processing for the drug development and can also provide better accuracy with the drugs being identified. Methods, such as supervised learning, involving classification and regression methods can help in the diagnosis of the diseases, drug efficacy and ADMET prediction (Guncar et al. 2018). Alternatively, unsupervised methods are useful in the discovery of disease subclasses through clustering techniques. The third category of algorithms, such as reinforcement learning, can be useful for predicting the de novo designing of the drugs (Chen et al. 2018). Thus, AI/ML can be highly useful as a tool for the identification of new compounds and repurpose the existing drugs.

1.2 Public Data Repositories

The list of public data repositories includes the following:

1.2.1 KAGGLE

Kaggle (www.kaggle.com) provides a large number of datasets, which is sufficient for the enthusiast to the expert. It supports the different types of file formats, which are very helpful for data publishing purpose, and they strongly inspire the dataset publishers to share their own data in an accessible, unpatented format. It provides an open-source, easy-to-use data layout that is better maintained through the platform and also provides datasets, which are effortless to operate together with more people, irrespective of their tools (Fig. 1.6).

It supports various file formats, which include CSV, Json, SQLite and archives.

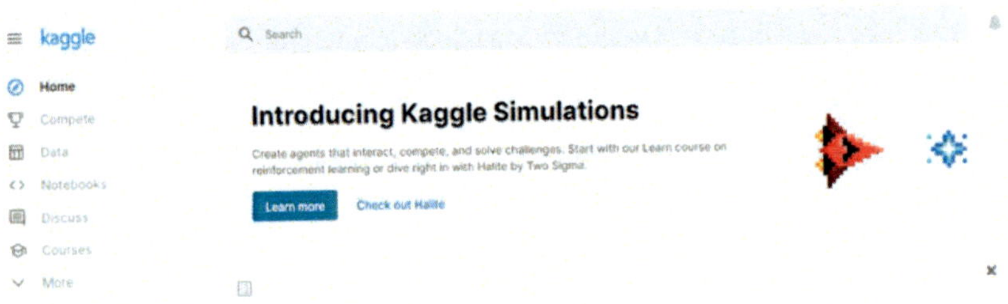

Fig. 1.6 Kaggle homepage. (www.kaggle.com)

1.2.2 Csv

The comma-separated list (CSV) is one of the most common file formats supported by the Kaggle. It is usually accessible for tabular data. CSVs uploaded in Kaggle should have field names on the header row in a readable format. On clicking "Data" tab of a dataset, a preview of the file's contents is visible in the data explorer. This makes it significantly easier to understand the contents of a dataset; an example is shown in Fig. 1.7, as there is no need to open the data in a notebook or download it.

CSV files will also have associated column descriptions and column metadata. The column descriptions allow you to assign descriptions to individual columns of the dataset, making it easier for users to understand what each column means.

1.2.3 JSON

JSON is also the most common file format for tree-like data that provides multiple layers, such as the branches on a tree. For example:

```
{[{'id': 0, 'type': 'bananas', 'quantity': 12}, {'id': 1, 'type': 'apples',
'quantity': 7}]}
```

For JSON files, the data tab will present an interactive tree with the nodes in the JSON file attached. You can click on the individual keys to open and disintegrate sections of the tree and can explore the structure of the dataset as you go along with it. JSON files do not support column descriptions or metrics.

1.2.4 SQLite

Kaggle supports database files in the form of lightweight SQLite format. SQLite databases consist of multiple tables, and each of it contains data in a tabular format. These tables support large datasets better than CSV files. The data tab represents each table in a database separately. The SQLite tables include column metadata and column metrics sections.

Fig. 1.7 An example to show the preview of the file's contents is visible in the data explorer by clicking on the data tab of dataset on Kaggle. (www.kaggle.com)

1.2.5 Archives

Archives are not a file format, but Kaggle also supports for files compressed using the ZIP file format as well as other common archive formats. These compressed files take up less space on the disk in comparison to uncompressed files, thus making them faster to upload to Kaggle and allowing you to upload datasets that will otherwise exceed the dataset size limitations. The archives do not populate with previews for individual file contents, but you can still browse the contents by the file name. The ZIP files and other archive formats can be the best choice for making image datasets available on Kaggle.

Datasets in Kaggle website is not a common Machine Learning (ML) dataset. Every dataset present in Kaggle consist of a community where people could discuss about data, discover new logic and methods from existing code and can create their own ML project using dataset in the Kaggle notebooks. We can find many different interesting datasets from all the field of all shapes and sizes. We can find the dataset through the data tab columns, newsfeed (if you logged in website) and tags and by searching the interested dataset from the search box (Fig. 1.8).

1.2.6 UCI ML Repository

The UCI Machine Learning Repository (archive.ics.uci.edu) consists of data repositories, data generation information and domain theories, which are accessed by the vast ML communities to analyse the different ML algorithms by doing experiments. In 1987, David Aha, along with his fellow students at the University of California, Irvine, created the first archive as ftp. Later on, the archive becomes very popular in all over the world and used by everyone. It is leading as a main resource for machine learning datasets (Fig. 1.9).

Fig. 1.8 Kaggle search box. (www.kaggle.com)

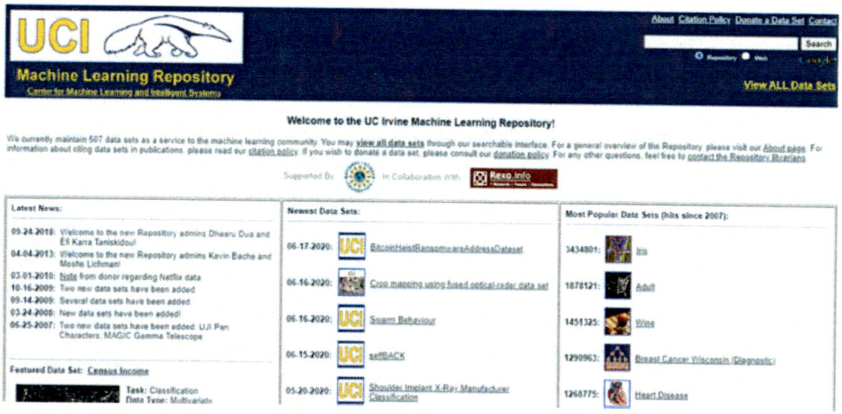

Fig. 1.9 UCI machine learning repository home page (including search box). (archive.ics.uci.edu)

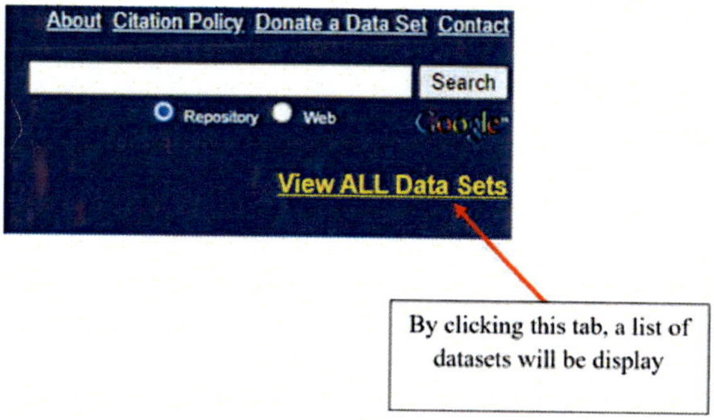

Fig. 1.10 Preview of "View ALL Datasets" tab. (archive.ics.uci.edu)

The archive has put up a great impact; till now, it's been cited about more than 1000 times and makes its presence in computer science field in one of the 100 most cited papers. By clicking the dataset description, tab users will be able to get the details about a particular dataset; they can even search for the desired dataset through the search box tab or by clicking on "View ALL Datasets" tab. The users even can download the datasets, which is divided into various categories, for example, according to the size of the dataset, or dataset can be used for a particular machine learning method. We can view all dataset present in UCI Machine Learning by clicking "View ALL Datasets" tab shown in Figs. 1.10 and 1.11. Currently, there are 507 datasets present in the repository.

For ease in searching the suitable dataset for AI/ML/DL task, the UCI Machine Learning Repository provides the columns "Browse Through:" shown in Fig. 1.11, in different sections such as "Default Task", "Attribute Type", 'Data Type", "Area", "Attributes" and "# Instances". These section helps to filter searching, so that we can get our interested dataset for our task. When we select dataset and click on it,

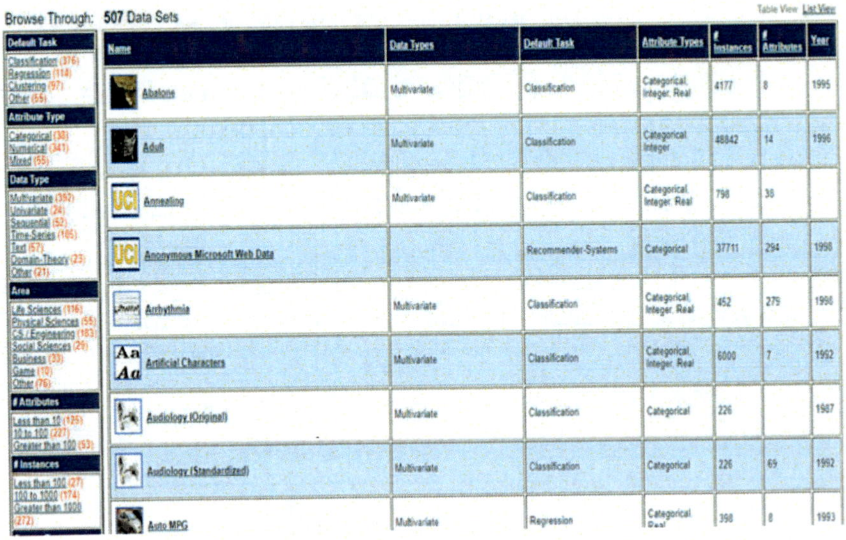

Fig. 1.11 List of datasets present in UCI Machine Learning Repository after clicking "View ALL Datasets" tab. (archive.ics.uci.edu)

repository will provide all the information and also include the case study about dataset in new window. From the "Data Folder" tab, we are able to download dataset file present in that directory. The "Data Description" tab provides a description about the dataset (Fig. 1.12).

1.2.7 HealthData.gov

HealthData.gov (catalog.data.gov) consists of datasets found across the American Federal Government with the aim of improving the health of American population (Fig. 1.13).

HealthData.gov provides a variety of datasets, such as environment related, public healthcare, medical instruments, medical aid, community service, chemical abuse and psychiatric health. The datasets are present in CSV, TXT, JSON, XSL and RDF file format.

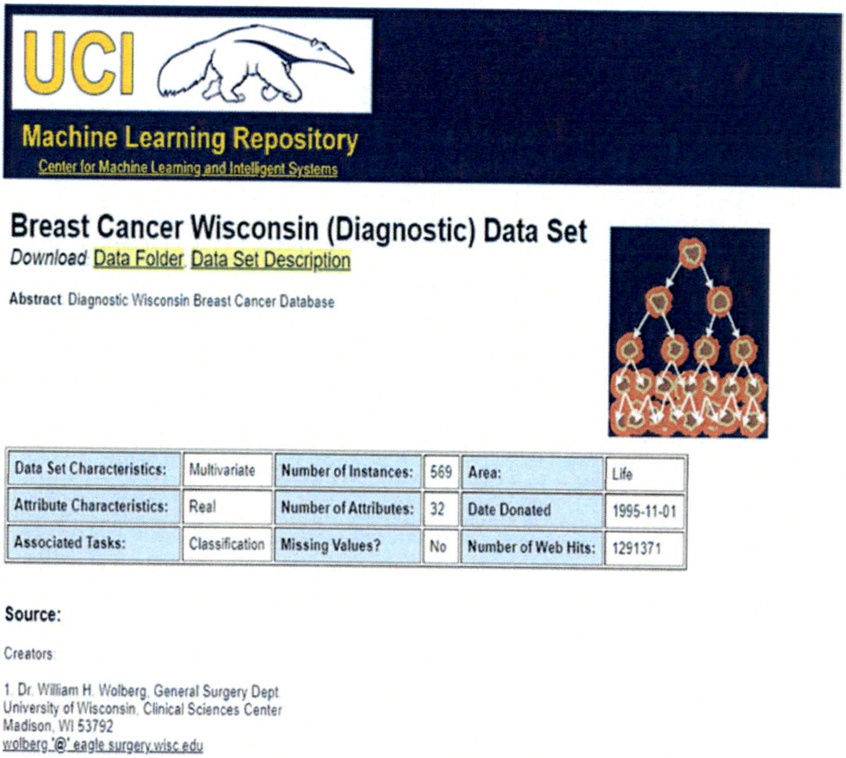

Fig. 1.12 Example of dataset window opened in UCI Machine Learning Repository. (archive.ics.uci.edu)

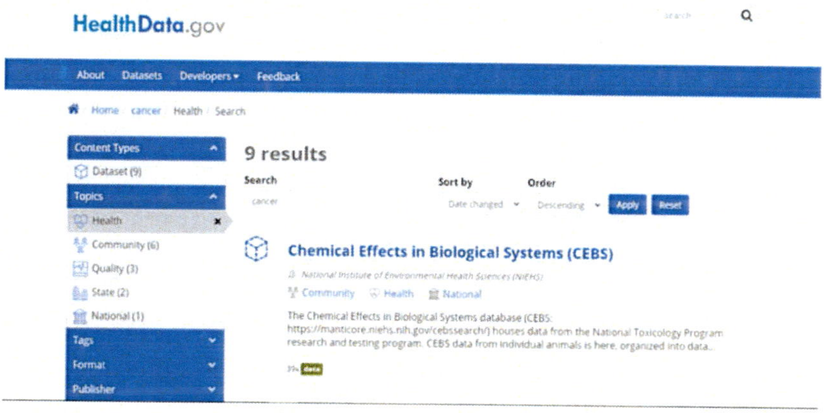

Fig. 1.13 HealthData.gov home page. (catalog.data.gov)

1.3 Review of Artificial Intelligence Techniques on Disease Data

1.3.1 Logistic Regression Model

Logistic regression model is the type of supervised learning technique that predicts the probability of a dependent qualitative variables. The logistic regression was termed as a function, which states an algorithm method known as the logistic function. It is a sigmoid function which forms a S-shaped curve, in which we can take any real value number that will be converted into a value that will be between 1 and 0, but that value will not be equal to 1 or 0 (Brownlee 2016).

$$1/\left(1+e^{-\text{value}}\right)$$

Here, e is represented as base of the natural log, and *-value* in the equation is represented as the real number value that is needed to be transformed. This function provides a plot that shows the numeric values between −5 and 5, in which logistic function is used to convert these values into the range of 1 and 0. The algorithm on the basis of logistic function works is called maximum likelihood estimation (MLE). MLE calculates the regression coefficient of the model that provides an accurate probability prediction of binary-dependent categorical/qualitative variables. The MLE algorithm works in an iterative process; thus, it will stop when the convergence criteria will meet. Therefore, any event will have its probability between 1 and 0.

Logistic regression is represented by an equation, which looks similar to linear regression equation. The input values/instances (x) are linearly combined to value of coefficient or we can refer to them as weights which predicts the value of an output (y) (Brownlee 2016).

The equation of logistic regression:

$$y=\frac{e^{(b0+b1*x)}}{\left(1+e^{(b0+b1*x)}\right)}$$

Here, y is represented as the output which is predicted, $b1$ is called coefficient of the input value x (single value) and $b0$ is the intercept. Input data in each column is associated with b coefficient (constant value), which can be obtained with the help of train dataset. The description of the model will be saved in a memory or file, which consists of coefficients.

The dependent categorical variable in the logistic regression model is called the binary variable, which contains an encoded numeric value 1 (e.g. positive) or 0 (e.-g. negative). The model predicts $p(y = 1)$, which is the function of x. Logistic regression algorithm fits the model that consists of binary classification data in an accurate manner by finding the best path for it. The logistic regression models are called to be members of generalized linear models. The logistic regression model predicts the probability of values between the range of 1 and 0. The prediction of probability through the logistic regression algorithm seems to be more accurate compared to other classifiers such as Naïve- Bayes, KNN, etc. The coefficients, which are formed by logistic model, provide a significance for each input value/ variable. The logistic regression models are mostly applicable, if the given data is categorical in nature, for example, cancer is malignant or not (1,0).

Logistic regression model is mostly used for classification task. It does not require to find any linear connection between dependent and independent variables. It is already able to manage different types of relations by using a non-linear log transformation to the predicted odds ratio. The model is useful in avoiding underfitting and overfitting. The logistic regression model needs large dataset, which is required for maximum likelihood estimation (MLE). It is difficult for the model to estimate MLE on small dataset.

There are three kinds of logistic regression model:

1. Binary logistic regression: Known as the categorical data, which contains two possible outcomes. For example: infected (1) and not infected (0).
2. Multinomial logistic regression: The categorical data, which contains more than two possible outcomes. For example: severe disease (1), mild disease (2) and no disease (0).
3. Ordinal logistic regression: When there are more than two categories in an order wise. For example: hospital facility rating from 1 to 5.

Logistic regression can be used to make prediction or prognosis, such as the risk of developing disease, for example, heart disease, cancer and diabetes, that can be done based on age, BMI, sex and anthropometric parameters (blood test results).

1.3.2 Artificial Neural Network Model

Artificial neural network (ANN) model is a subset of deep learning (DL) technique, which is build using a vast number of elements known as neurons. Every neuron will make a decision and then transfers this decision to other neurons that are arranged like an interconnected layer. The artificial neural network (ANN) model can imitate any task and try to generate answer to any practical question, with the help of a large amount of training dataset and computation strength. The artificial neural network has only three layers:

1. Input layer: It takes input values or non-dependent variables for building a model.
2. Hidden layer.
3. Output layer: It generates predictions.

The process of artificial neural network (ANN) model is shown in the flow chart given in Table 1.1 (Fig. 1.14).

All the linkages present in the artificial neural network (ANN) model has the same calculation. A sigmoid relation is presumed between the input nodes and the

Table 1.1 Flow chart of ANN process

Random allocation of weights to all the edges
⬇
By employing the linkage and the inputs between both the input and hidden node, we activation rate of the hidden node is calculated.
⬇
Activation-rate of the output nodes is calculated activation- rate of hidden nodes and linkages output nodes
⬇
Error-rate of the output nodes is calculated and links are adjusted between hidden and output nodes
⬇
calculated weights and error at output node will be used to drop the error to hidden nodes
⬇
Adjustment between hidden nodes and inputs nodes weights is done
⬇
This procedure will be repeated in an iterative manner, till the convergence criteria are met
⬇
Scores of last linkages weight the activation-rate of the output nodes.

rate of activation of hidden nodes (Srivastava 2014). An equation is shown below to calculate the activation rate of *H1*:

$$\text{Logit}(H1) = W(I1 * H1) * I1 + W(I2 * HI) * I2 + W(I3HI) * I3 + \text{Constant} = > P(HI)$$

Fig. 1.14 Artificial neural network model

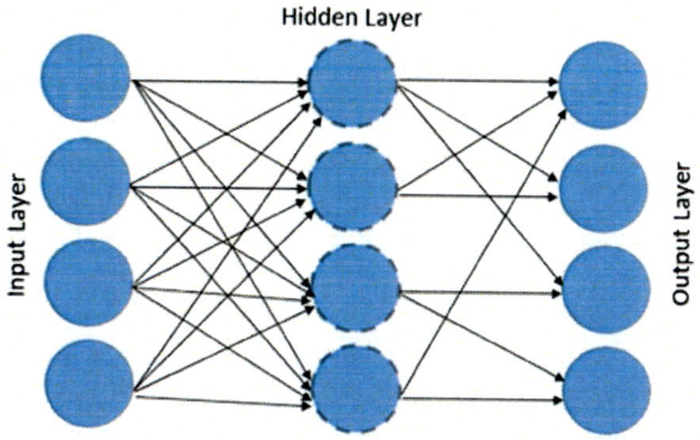

$$= \frac{1}{(1 + e^{(-f.)})}$$

The artificial neural network (ANN) algorithm helps to understand the increase/decrease of dataset impact and understand the situations where the model fits the best. The artificial neural network (ANN) model is very useful in disease management-related problem, such as disease diagnosis, cancer prediction, speech recognition, duration of disease prediction (HIV-AIDS) (Park and Chang 2001), image prediction analysis and its interpretation. For example: an automated electrocardiographic (ECG) was implemented, which was useful in the diagnosis of myocardial infarction (Bartosch-Härlid et al. 2018) and drug development. It is also applicable in non-clinical problems that include improvement in the organizational management in healthcare fields (Goss and Vozikis 2002), predicting the key indicators, such as the cost price or utilization of facilities (Kaur and Wasan 2006). Artificial neural network (ANN) model usually is used as a decision supporting model that helps the healthcare suppliers and the healthcare system with a cost-effective solution to time and resource handling (Nolting 2006).

1.3.3 Support Vector Machine Model

Support vector machine (SVM) model is a type of supervised ML technique. It is used in classification- and regression-related problems but, generally, applies in classification analysis purposes. The SVM can deal with categorical and continuous variables. SVM model shows the portray of various classes in a hyperplane in multidimensional space. The generation of hyperplane in iterative manner by SVM in order minimize the error (Ray 2017). The main objective of SVM is to classify the datapoints of dataset based on a maximum marginal hyperplane.

There are some important terms in SVM model that need to be known:

- *Support vectors*: The support vectors are called datapoints, which are located near to the optimal hyperplane. Support vector helps to define the separating line.
- *Hyperplane*: Hyperplane helps to divide the dataset into classes.
- *Margin*: Margin tells the distance between the datapoints from different classes based on support vectors.
- *SVM kernels*: The SVM kernels help to separate the non-separable datapoints efficiently by adding more dimension to it. The types of kernel are:
  - Linear
  - Polynomial
  - RBF In SVM model process, the first aim is to identify the points from the given two classes, which is nearest to the hyperplane. These identified points are called support vectors. After that, the model will calculate the distance between the hyperplane and the support vectors shown in Fig. 1.11. This distance between them is called margin. The main goal is to increase the margin, and for that, process goes on iteratively. The hyperplane, which has the highest marginal rate, is called suitable separating line for dividing the two classes (Pupale 2018). SVM model builds the decision-based separating line in such a precise manner, in order to have a division between the two classes as wider as possible. The SVM model works well in high-dimensional spaces. Its relative memory is efficient. The SVM model is very useful in situations such as when dimensions of the data are higher compared to instances of number. The SVM model is very useful in predictive modelling, such as in the diagnosis/prognosis of disease (e.g. breast cancer) (Patrício et al. 2018), identifying and classifying the genes and patients on the basis of genes or other biological problems. SVM modelling is called to be an optimistic approach for predicting medication adherence in heart failure patients (Lee et al. 2010). A device e-doctor is a web-based application that makes an automated diagnosis about health-related problems (Karakülah et al. 2014). The device was built on SVM model/algorithm that analyses the data and then proceed to decisions, based on their knowledge. With the help of EHR record data, the SVM model is able to understand each health-related problem that can be diagnosed by the device (Kampouraki et al. 2013).

1.4 Case Study: Parkinson's Disease Prediction

Parkinson's disease is called a neurological disease, which causes stiffness and shakiness in the body and difficulty in walking, balancing and coordination. The signs of Parkinson's generally begin in a slow manner, but later on, it gradually becomes adverse. When disease progression happens, the Parkinson patients start facing struggle in walking and talking. Even the affected people are faced with mental illness and mood swings problems, insomnia, difficulty in memorizing things and fatigue. The men and women both are affected by Parkinson's disease. But, around more than 50 per cent of men are affected by this disease compared to women. The main element of danger for this neurological disease is age factor. As in many cases, people suffering from Parkinson's have their early-stage encounter with the disease at the age of 60; about more than 5 per cent of people suffering from Parkinson's disease encounter with early stage of disease that may begin at their 40s. The outset of this disease is frequent; it may be inheriting, or some forms could be linked to a specific gene mutation (Michelle 2019).

Parkinson's disease impaired the nerve cells, which is an important part of the brain that maintains and controls all the movement within the brain. A chemical called dopamine is produced by the nerve cells, which is important for brain working. But due to Parkinson's, these nerve cells get damaged, and when dopamine production gets low, it causes difficulty in movement. The researchers still do not know what are the reasons that lead to the death of the nerve cells that produce dopamine. Parkinson's also affects the patient's the nerve cell terminal, which produces the chemical known as norepinephrine; it is a chemical messenger, which is crucial for sympathetic nervous system, that helps to supervise various involuntary body functions, such as breathing, heart beating, blood pressure and reflexes. The loss of norepinephrine may cause panic attacks, stress, fatigues, high blood pressure, depression difficulty in digesting food, hypotension, etc.

The symptoms of Parkinson's disease are:

- Difficulty in balancing and coordination, which can cause falls.
- Stiffness of the limbs and trunk.
- Slow motion.
- Tremor in the hands, legs and heads.

Some other symptoms are depression, mood swings, difficulty in eating and speaking, constipation and sleep disruptions. There are still gaps in treating Parkinson's disease. There is no confirmed medical test is that can surely reveal Parkinson's disease. Thus, it causes difficulty in diagnosing the disease accurately. Even after diagnosis, there is still no such confirmed remedy for Parkinson's disease. There are some medications used to manage the disease but still not very much effective.

The artificial intelligence (AI) and machine learning (ML) are emerged as a new weapon that helps to fight with Parkinson's disease. The AI/ML techniques help clinicians/neurologists to diagnose the disease and understand the disease prognosis. Artificial intelligence (AI) and neurological disorders in combination will help researchers to have strong insights on disease progression, so that they will be able to develop full proof plan for more effective treatments than performing the traditional medical diagnosis treatment that take lot of time to reveal. With the help of AI, the costs related to treatment and healthcare system will be reduced. In a study, a model was proposed on thalamocortical dysrhythmia (TCD) that was used to give a brief detail on the different types of neurological diseases. It was distinguished through an oscillatory pattern, in which the resting-state alpha activity was taken over from cross-frequency coupling of high- and low-frequency oscillations (Vanneste et al. 2018). Support vector machine (SVM) learning was used as a data-driven approach for analysing the oscillatory patterns of resting-state electroencephalography in the person suffering from Parkinson's disease, depression, tinnitus and neuropathy. Artificial intelligence (AI) helps clinicians to distinguish Parkinson's disease patients from healthy people and tries to discover the different characteristics that are associated with Parkinson's disease (Rehme et al. 2015). Artificial Intelligence (AI) technology, powered with cloud-based digital platform has been created that helps to differ the Parkinson's Disease patients from healthy person (Tsoulos et al. 2019).

In this case study, we will be demonstrating the practical applications of AI- and ML-based techniques on Parkinson's disease data that will be retrieved from public repositories from Kaggle. Further, the process of importing and preprocessing the dataset on Parkinson's disease will be shown. Finally, we will elaborate the process of building a predictive model using different classifier on retrieved Parkinson's disease dataset using MATLAB.

MATLAB is a high-level language in the technical computation. It combines programming, computing and visualization all together. It provides a convenient environment when problem is present with mathematical notation. MATLAB means matrix laboratory, which was first created to do matrix computation easily. MATLAB is the system that provides the data element as array; there is no need of doing dimensioning of data. It helps to solve technical computation easily, especially vector and matrix. It takes less time to write the program in non-interactive form such as FORTRAN and C language. MATLAB has been developed so much nowadays through the inputs given by MATLAB community. In the university platform, MATLAB is used as an interactive tool for professional courses in science and engineering. In industries, MATLAB is for R&D purposes and data analysis. MATLAB provides a lot of features, and one of the important features is it provides the different toolboxes that are used for specific solutions. These toolboxes helps users to understand, learn and build applications towards specialized technology. These toolboxes provide inclusive collections of MATLAB functions (as M-file format) that has more expand the MATLAB language and makes it able to do any classes of problems. There are some fields on which toolbox is available such as bioinformatics, machine learning and statistics, neural networks, etc. Artificial intelligence (AI) or machine learning (ML) become easier with the help of MATLAB language. MATLAB provides beneficial machine learning functions; thus, there is no need to do complicated maths that are required in machine learning stuffs.

1.4.1 Importing the Data

The Parkinson's disease dataset is retrieved from Kaggle.com (www.kaggle.com/wajidsaw/detection-of-parkinson-disease).

The dataset is in CSV file format, consisting of 23 attributes and 195 instances. In that 23 attributes, 22 attributes are features, sound data of Parkinson's disease patients and healthy people, and the last attribute is label class, which consists of

```
%% --------------- Importing the dataset -------------------------
% ---------------------------- Code -------------------------
data = readtable('datasets_410614_786211_parkinsons.csv');
```

Fig. 1.15 Code for importing the Parkinson's disease data

```
COMMAND WINDOW

data =

  195×23 table
```

Var1	Var2	Var3	Var4	Var5	Var6	Var7	Var8	Var9	Var10	Var11	Var12	Var13	Var14	Var15	Var16
119.99	157.3	74.997	0.00784	7e-05	0.0037	0.00554	0.01109	0.04374	0.426	0.02182	0.0313	0.02971	0.06545	0.02211	21.033
122.4	148.65	113.82	0.00968	8e-05	0.00465	0.00696	0.01394	0.06134	0.626	0.03134	0.04518	0.04368	0.09403	0.01929	19.085
116.68	131.11	111.56	0.0105	9e-05	0.00544	0.00781	0.01633	0.05233	0.482	0.02757	0.03858	0.0359	0.0827	0.01309	20.651
116.68	137.87	111.37	0.00997	9e-05	0.00502	0.00698	0.01505	0.05492	0.517	0.02924	0.04005	0.03772	0.08771	0.01353	20.644
116.01	141.78	110.66	0.01284	0.00011	0.00655	0.00908	0.01966	0.06425	0.584	0.0349	0.04825	0.04465	0.1047	0.01767	19.649
120.55	131.16	113.79	0.00968	8e-05	0.00463	0.0075	0.01388	0.04701	0.456	0.02328	0.03526	0.03243	0.06985	0.01222	21.378
120.27	137.24	114.82	0.00333	3e-05	0.00155	0.00202	0.00466	0.01608	0.14	0.00779	0.00937	0.01351	0.02337	0.00607	24.886
107.33	113.84	104.31	0.0029	3e-05	0.00144	0.00182	0.00431	0.01567	0.134	0.00829	0.00946	0.01256	0.02487	0.00344	26.892

Fig. 1.16 Parkinson's disease dataset imported in MATLAB

```
%%----------------Data Preprocessing --------------------------
%%------------- Checking Missing Value--------------------------
ismissing(data)
```

Fig. 1.17 Code for checking missing value in dataset

two classes: class 1, Parkinson's disease patient [1], and class 2, healthy person [0]. After retrieving, we will import the Parkinson's disease data in MATLAB editor script using "readtable ( )" function (Fig. 1.15).

By running this code, the given dataset will be imported in MATLAB, and our dataset will be seen in the command window of MATLAB shown in Fig. 1.16. The dataset looks like this in MATLAB.

1.4.2 Data Preprocessing and Feature Selection

Data preprocessing is the process that is used to prepare/format the raw data, making it suitable for building machine learning model. It is a one of the major processes required for building ML model. The data preprocessing step includes the following: checking missing value, categorical data dealing and feature scaling (standardization or normalization).

Checking Missing Value in Dataset

To check/find any missing values in the dataset, we can use "ismissing ( )", a MATLAB function (Fig. 1.17).

```
195x23 logical array

0 0 0 0 0 0 0 0 0 0 0 0 0 0 0 0 0 0 0 0 0 0 0
0 0 0 0 0 0 0 0 0 0 0 0 0 0 0 0 0 0 0 0 0 0 0
0 0 0 0 0 0 0 0 0 0 0 0 0 0 0 0 0 0 0 0 0 0 0
0 0 0 0 0 0 0 0 0 0 0 0 0 0 0 0 0 0 0 0 0 0 0
0 0 0 0 0 0 0 0 0 0 0 0 0 0 0 0 0 0 0 0 0 0 0
0 0 0 0 0 0 0 0 0 0 0 0 0 0 0 0 0 0 0 0 0 0 0
0 0 0 0 0 0 0 0 0 0 0 0 0 0 0 0 0 0 0 0 0 0 0
0 0 0 0 0 0 0 0 0 0 0 0 0 0 0 0 0 0 0 0 0 0 0
0 0 0 0 0 0 0 0 0 0 0 0 0 0 0 0 0 0 0 0 0 0 0
0 0 0 0 0 0 0 0 0 0 0 0 0 0 0 0 0 0 0 0 0 0 0
0 0 0 0 0 0 0 0 0 0 0 0 0 0 0 0 0 0 0 0 0 0 0
0 0 0 0 0 0 0 0 0 0 0 0 0 0 0 0 0 0 0 0 0 0 0
0 0 0 0 0 0 0 0 0 0 0 0 0 0 0 0 0 0 0 0 0 0 0
0 0 0 0 0 0 0 0 0 0 0 0 0 0 0 0 0 0 0 0 0 0 0
```

Fig. 1.18 Output of missing value code

```
%%---------------Data Preprocessing -----------------------------
%%------------- Checking outlier---------------------------
isoutlier(data)
```

Fig. 1.19 Code for outlier detection

After running this code, a logical table will form the dataset in the command window, where “0” is denoted as “no missing value” and “1” is denoted as “missing value”. The output of our dataset doesn’t show any missing value, which is shown in Fig. 1.18.

Like this, we can identify the missing value in a dataset easily.

Dealing with Categorical Data

Our dataset doesn’t contain any categorical data (yes or no); it is numeric or real-value dataset. Thus, there is no need to do this step for a given dataset.

Outlier Detection

An outlier is called an instance, which diverges from an overall pattern on a sample that could affect in implementing the machine learning model. On MATLAB, we can analyse and detect outlier using MATLAB function “isoutlier( )”, which helps to identify the outlier in dataset; if there is outlier in the dataset, we can remove it is using “rmoutlier(data, method)” method = mean/median/quartiles/grubbs/gesd. We will check outlier on the dataset and the code will again give the logical table, where “0” is denoted as no outlier and “1” is denoted as outlier, and if there is any, we will remove it (Fig. 1.19).

So, like this, we can detect and analyse the outlier in the dataset.

```
%%--------------Data Preprocessing --------------------------
%% ------------- Feature Scalling ----------------------------
stand_var15 = (data.Var15 - mean(data.Var15))/std(data.Var15);
data.Var15 = stand_var15;

stand_var16 = (data.Var16 - mean(data.Var16))/std(data.Var16);
data.Var16 = stand_var16;


stand_var17 = (data.Var17 - mean(data.Var17))/std(data.Var17);
data.Var17 = stand_var17;


stand_var18 = (data.Var18 - mean(data.Var18))/std(data.Var18);
data.Var18 = stand_var18;

stand_var19 = (data.Var19 - mean(data.Var19))/std(data.Var19);
data.Var19 = stand_var19;

stand_var20 = (data.Var20 - mean(data.Var20))/std(data.Var20);
data.Var20 = stand_var20;

stand_var21 = (data.Var21 - mean(data.Var21))/std(data.Var21);
data.Var8 = stand_var8;

stand_var22 = (data.Var22 - mean(data.Var22))/std(data.Var22);
data.Var22 = stand_var22;
```

Fig. 1.20 Feature scaling code

Feature Scaling

Feature scaling is a method that standardizes the features present in the dataset into a given fixed range. The feature scaling step, using the standardisation method, is needed on the given dataset, because the given dataset contains datapoints with different ranges. Thus, it needs to be in standardized range; otherwise, it will create problems in building classifier, as many classifiers (such as SVM) are sensitive to ranges of datapoints. The standardization formulae will use a code for feature scaling step on MATLAB. The formulae of standardization are given below:

$$X_{\text{new}} = X_i - X_{\text{mean}}/\text{Standard Deviation}$$

We will apply feature scaling step on the features of the dataset, not on the label class (Fig. 1.20).

After feature scaled step, the given dataset gets scaled. The data preprocessing part is completed. Now, we will do the feature selection from the given Parkinson's disease dataset that helps to reduce the overfitting problems and can provide good accuracy of machine learning (ML) model. For feature selection, we will be using the principal component analysis (PCA) algorithm on a given dataset. PCA is a ML method/technique that helps to reduce the dimensions of multivariate datasets

```
%%--------------- Dimensionality Reduction --------------------
%% -------------------------- PCA -------------------------------
% -------------------------- Code ----------------------------
class_labels = data.Var23;
data = table2array(data(:,1:end-1));
[coeff,score,latent,tsquared,explained,mu] = pca(data);
Var1 = score(:,1);
Var2 = score(:,2);
data = table(Var1, Var2, class_labels)
```

Fig. 1.21 Feature selection code

(decreases overfitting problems), and it increased interpretation without loss of information. As the given dataset is high-dimensional (22 features), it may cause difficulty in projection. The PCA technique will reduce the dimension, by finding the new variables set that will be smaller than the original variable set, but it will contain most of the dataset information. The PCA is done by calculating the covariance/correlation matrix of the original dataset and then performs the eigenvalue decomposition. Eigenvalue decomposition means the computation of eigenvalue and eigenvector. The eigenvectors (principal components) represent the direction of new variables, while eigenvalues represent the magnitude of the new variables, and the variance percentage corresponds to the principal component on the basis of which new variable set is selected.

The pca ( ) is MATLAB function that helps to do PCA analysis as shown in Fig. 1.21. By running this function, it will return different parameters such as coeff, score, latent, tsquared, explained, mu computed by pca ( ) function. Our main interest is with two parameters, that is, explained and score parameters.

The output of the explained parameter tells us the total variance percentage (eigenvalues) that is explained by principal components(eigenvectors). Therefore, in total, there are 22 principal components (eigenvectors), which are obtained with their corresponding variance percentage (eigenvalues). Explained variance graph result shows that principal components 1 and 2 both make up approximately 76% of the total variance (covers 76% of data information from the original dataset) and the rest principal component variance percentage decreases gradually and, thus, makes these first two elements ideal as a new variable set. For better explanation, graphical representation of explained variance is also shown in Fig. 1.22.

Then, we will create a new table, in which Var 1 and Var 2 scores will be instances, or datapoints which help to make classification predictions along with class_labels attribute.

This new table of data from the original dataset have two new attributes/variables, that is, Var 1 and Var 2 (obtain from PCA analysis), and the third attribute is the class_labels shown in Fig. 1.23.

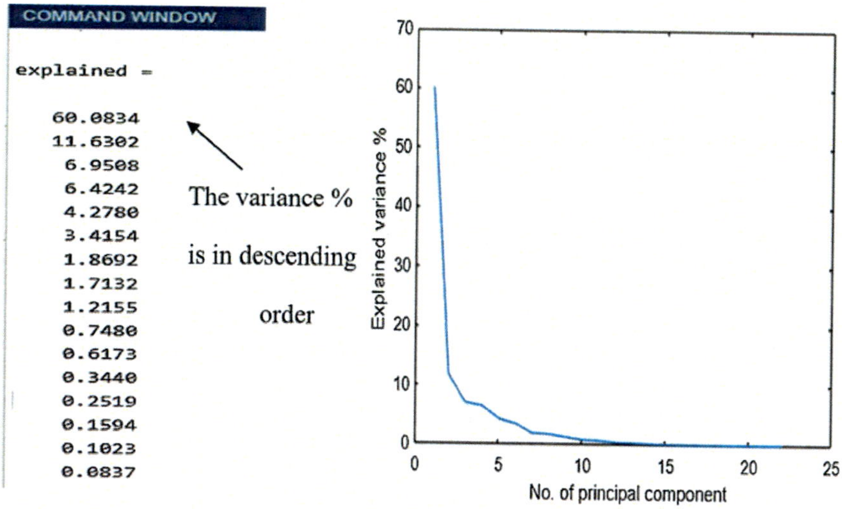

Fig. 1.22 Output of explained variance percentage along with graphical representation

1.4.3 Building Classifier

Using MATLAB function fitc, we can perform classification using a different classifier, such as KNN, SVM and Naive Bayes. By running this function, the classifier/model will learn/train from input data that have labels (class_labels) for predictive modelling. You can build each classifier one by one or together by changing the variable name (Fig. 1.24).

Output of these classifiers (Fig. 1.25):

1.4.4 Predictive Modelling

In predictive modelling part, we will first divide the given dataset in train and test set. The ideal ratio for the division of dataset is 80:20, 60:40 and 70:30. Here, we will divide the given dataset into 60:40 ratio as train (60%) and test/validation (40%) set. The cvpartition( ) MATLAB function helps to do random partition on set of data to a specific size. The holdout method does the partition of data exactly into two part or subset for training and validation (Fig. 1.26).

The given dataset has 195 instances; therefore, according to 60:40 ratio, the above code will divide the dataset of 117 instances as train size and 78 as test/validation size shown in Fig. 1.27.

After train and test division, will we now train our classifier/model (SVM, KNN and Naive Bayes) on train set (consists of 117 instances). Crossval ( ) function helps to cross-validate the classification model, which means, it helps to train the model on train set, and this is done by putting cv in code. We train the models/classifiers only one time (Fig. 1.28).

```
data =

  195x3 table
```

Var1	Var2	class_la
2.1496	-1.4482	1
4.7146	-1.2566	1
3.9066	-1.2807	1
4.1665	-1.4742	1
5.7641	-0.93434	1
3.1778	-1.4488	1
-1.8077	-1.9052	1
-2.2596	-1.9619	1
-0.17875	-2.5164	1
0.47309	-2.6242	1
-2.1946	-1.1639	1
-2.0438	-1.7038	1
-2.9454	-1.0556	1
2.4789	-1.7182	1
1.4941	-1.4859	1
2.02	-1.5842	1
2.7452	-1.3492	1
2.4912	-0.99077	1
-0.29304	-1.1183	1
1.0048	-0.97044	1
3.9905	1.4017	1
0.063919	0.65498	1
-1.4328	0.61906	1
-1.7386	1.5456	1
-0.7974	0.73537	1
9.0522	0.41329	1
6.0869	1.2446	1
8.1981	1.3721	1
6.7484	4.0099	1
-2.0691	-0.67584	1
-1.6223	0.28907	1
-1.7562	-0.43114	1
-2.2632	0.41321	1
-1.3235	-1.2756	0
-0.72052	-1.2128	0
-2.0689	-1.0899	0
-1.8709	0.92376	0
-2.9482	0.84066	0
-2.872	-0.97798	0
-0.55406	1.5248	0
0.025175	1.4558	0
-0.72436	1.4764	0
0.9043	1.9128	0
-0.5457	2.1251	0
-1.2335	1.1561	0

Fig. 1.23 New table of dataset (after PCA)

```
%% -------------- Building Classifier -----------------------------
% ---------------------------- Code ----------------------------

classification_model = fitcsvm(data,'class_labels');
classification_model = fitcknn(data,'class_labels');
classification_model = fitcnb(data,'class_labels');
```

Fig. 1.24 Building classifier (SVM, KNN and Naive Bayes) code

```
ClassificationSVM
            PredictorNames: {'Var1'  'Var2'}
              ResponseName: 'class_labels'
     CategoricalPredictors: []
                ClassNames: [0 1]
            ScoreTransform: 'none'
           NumObservations: 195
                     Alpha: [76×1 double]
                      Bias: 1.3611
          KernelParameters: [1×1 struct]
            BoxConstraints: [195×1 double]
           ConvergenceInfo: [1×1 struct]
           IsSupportVector: [195×1 logical]
                    Solver: 'SMO'

Properties, Methods
```

```
ClassificationKNN
           PredictorNames: {'Var1'  'Var2'}
             ResponseName: 'class_labels'
    CategoricalPredictors: []
               ClassNames: [0 1]
           ScoreTransform: 'none'
          NumObservations: 195
                 Distance: 'euclidean'
             NumNeighbors: 1

Properties, Methods
```

```
ClassificationNaiveBayes
              PredictorNames: {'Var1'  'Var2'}
                ResponseName: 'class_labels'
       CategoricalPredictors: []
                  ClassNames: [0 1]
              ScoreTransform: 'none'
             NumObservations: 195
           DistributionNames: {'normal'  'normal'}
      DistributionParameters: {2×2 cell}

Properties, Methods
```

Fig. 1.25 Output of different classifiers

```
%% --------------- Test and Train sets ----------------------------
% ------------------------------ Code ----------------------------
cv = cvpartition(classification_model.NumObservations, 'HoldOut', 0.4);
```

Fig. 1.26 Code for dividing the dataset into training and testing set

```
cv =

  Hold-out cross validation partition
     NumObservations: 195
         NumTestSets: 1
           TrainSize: 117
            TestSize: 78
```

Fig. 1.27 Output of train and test size of dataset

```
%% -------------Training the model/classifier ----------------------------
% ---------------------------- Code ----------------------------
cross_validated_model = crossval(classification_model,'cvpartition',cv);
```

Fig. 1.28 Code to train a model

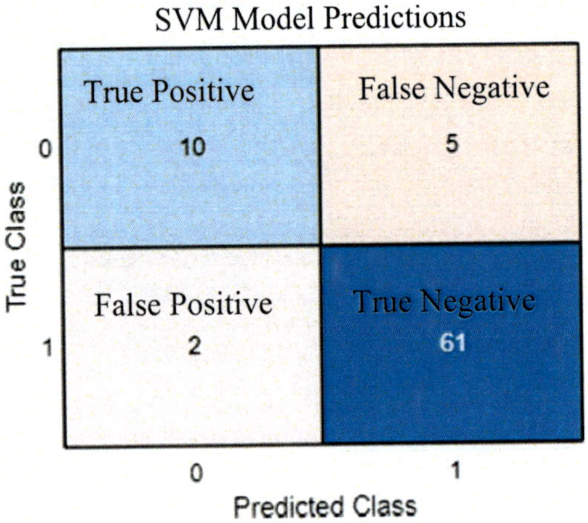

Fig. 1.29 Confusion matrix of SVM model

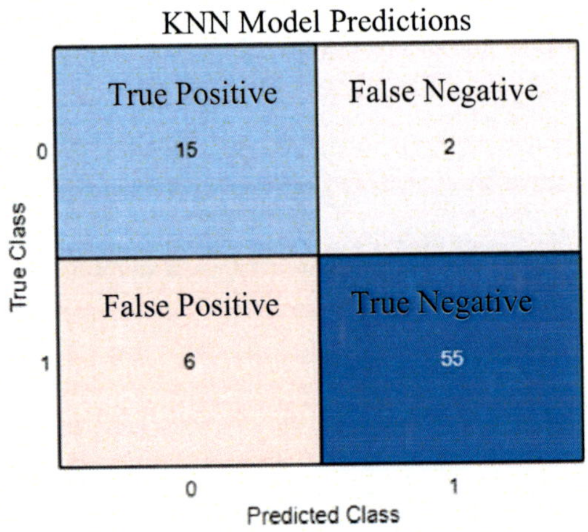

Fig. 1.30 Confusion matrix of KNN model

1.4.5 Performance Validation of the Model

In the performance validation part, the trained model/classifier (SVM, KNN and Naïve Bayes) will make prediction on test or validation or unseen data (78 instances), called performance validation. The prediction made by the predictive model will

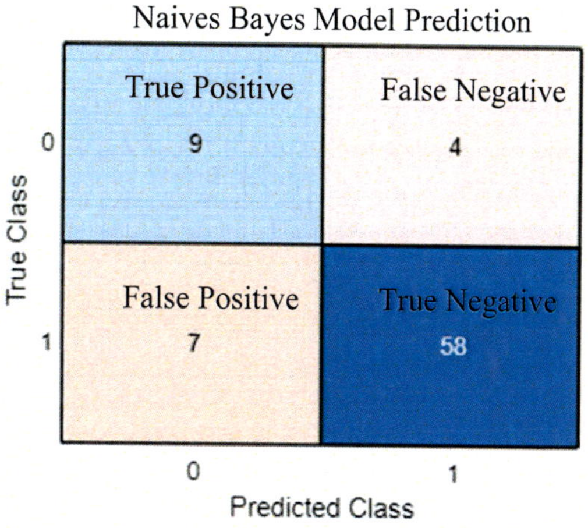

Fig. 1.31 Confusion matrix of Naive Bayes model

show results in a form of confusion matrix. Confusion matrix helps to know the performance of a predictive model. It shows the ways, in which a predictive model gets confused in making predictions. The correct and incorrect prediction numbers are sum up with values and divided into each class. The prediction is made by SVM, KNN and Naive Bayes models, and confusion matrix is shown in Figs. 1.29, 1.30 and 1.31:

In the above figures, there are two classes: "0" is denoted as "healthy person" and "1" is denoted as "Parkinson's disease patient". The other terms mean:

- True positives (TP): These are the instances which the model predicted as "0" (healthy person), and in an actual case also, they are healthy persons.
- True negatives (TN): These are the instances which the model predicted as "1" (Parkinson's disease patient), and in an actual case also, they are Parkinson's disease patients.
- False positives (FP): These are the instances which the model predicted as "0" (Healthy Person), but in an actual case, they are Parkinson's disease patient. This type of error is also called a "Type I error."
- False Negatives (FN): The instances which model predicted as "1" (Parkinson's disease patient), but in actual case, they are healthy persons. This type of error is also called a "Type II error".

Therefore, the left-side diagonal of confusion matrix portrays as "correct predictions" and right- side diagonal of confusion matrix portrays as "incorrect predictions". The classification rate or accuracy of all three models for making correct predictions are:

- SVM model accuracy: 91%.
- KNN model accuracy: 89.74%.
- Naive Bayes model accuracy: 85.89%.

Thus, further analysis on predictive model can be done in the future.

References

Akella B (2020) Types of machine learning – supervised and unsupervised learning. https://intellipaat.com/blog/tutorial/machine-learning-tutorial/types-of-machine-learning/. Accessed 5 July 2020

Alanine A, Nettekoven M, Roberts E et al (2003) Lead generation - enhancing the success of drug discovery by investing in the hit to Lead process. Comb Chem High Throughput Screen 6 (1):51–66. https://doi.org/10.2174/1386207033329823

Anderson AC (2011) Structure-based functional design of drugs: from target to lead compound. Methods Mol Biol 823:359–366. https://doi.org/10.1007/978-1-60327-216-2_23

Bakkar N, Kovalik T, Lorenzini I (2018) Artificial intelligence in neurodegenerative disease research: use of IBM Watson to identify additional RNA-binding proteins altered in amyotrophic lateral sclerosis. Acta Neuropathol 135:227–247. https://doi.org/10.1007/s00401-017-1785-8

Bartosch-Härlid A, Andersson B, Aho U et al (2018) Artificial neural networks in pancreatic disease. Br J Surg 95(7):817–826. https://doi.org/10.1002/bjs.6239

Belić M, Bobić V, Badža M et al (2019) Artificial intelligence for assisting diagnostics and assessment of Parkinson's disease—a review. Clin Neurol Neurosurg 184:105442. https://doi.org/10.1016/j.clineuro.2019.105442

Brownlee J (2016) Learning, logistic regression for machine. machinelearningmastery.com. Accessed 5 July 2020

Chen H, Engkvist O, Wang Y et al (2018) The rise of deep learning in drug discovery. Drug Discov Today 23(6):1241–1250. https://doi.org/10.1016/j.drudis.2018.01.039

Colangelo M (2020) AI driven biomarkers could help prevent age-related diseases. https://www.forbes.com/sites/cognitiveworld/2020/01/28/ai-driven-biomarkers-of-aging/#6f2bde37c94f. Accessed 3 July 2020

Croft P, Altman DG, Deeks JJ et al (2015) The science of clinical practice: disease diagnosis or patient prognosis? Evidence about "what is likely to happen" should shape clinical practice. BMC Med 13:20. https://doi.org/10.1186/s12916-014-0265-4

Datta S, Barua R, Das J (2019) Application of artificial intelligence in modern healthcare system. In: Pereira L (ed) Alginates, Chapter 8. IntechOpen, Rijeka. https://doi.org/10.5772/Intechopen.90454

Dilsizian SE, Siegel EL (2014) Artificial intelligence in medicine and cardiac imaging: harnessing big data and advanced computing to provide personalized medical diagnosis and treatment. Curr Cardiol Rep 16:441. https://doi.org/10.1007/s11886-013-0441-8

Ferrero E, Dunham I, Sanseau P (2017) In silico prediction of novel therapeutic targets using gene–disease association data in silico prediction of novel therapeutic targets using gene–disease association data. J Transl Med 15:182

George L. (2020) AI & medicine releases biomarker discovery and targeted proteomics services for researchers. www.clinicalresearchnewsonline.com. Accessed 27 July 2020

Goss EP, Vozikis GS (2002) Improving health care organizational management through neural network learning. Health Care Manag Sci 5:221–227. https://doi.org/10.1023/A:1019760901191

Guncar G, Kukar M, Notar M et al (2018) An application of machine learning to haematological diagnosis. Sci Rep 8:411. https://doi.org/10.1038/s41598-017-18564-8

Huang S, Yang J, Fong S et al (2020) Artificial intelligence in cancer diagnosis and prognosis: opportunities and challenges. Cancer Lett 471:61–71. https://doi.org/10.1016/j.canlet.2019.12.007

Kampouraki A, Vassis D, Blesis P et al (2013) e-Doctor: a web based support vector machine for automatic medical diagnosis. Procedia Soc Behav Sci 73:467–474. https://doi.org/10.1016/j.sbspro.2013.02.078

Karakülah G, Dicle O, Koşaner O et al (2014) Computer based extraction of phenoptypic features of human congenital anomalies from the digital literature with natural language processing

techniques. Stud Health Technol Inform 205:570–574. https://doi.org/10.3233/978-1-61499-432-9-570
Kaur H, Wasan SK (2006) Empirical study on applications of data mining techniques in healthcare. J Comput Sci 2(2):194–200. https://doi.org/10.3844/JCSSP.2006.194.200
Ko J, Baldassano SN, Loh PL et al (2019) Machine learning to detect signatures of disease in liquid biopsies - a user's guide. Lab Chip 18(3):395–405. https://doi.org/10.1039/C7LC00955K
Lee S, Son YJ, Kim J et al (2010) Healthcare. Inf Res 16(4):253–259. https://doi.org/10.4258/hir.2014.20.2.125
Lee JG, Jun S, Cho YW et al (2017) Deep learning in medical imaging: general overview. Korean J Radio 18(4):570–584. https://doi.org/10.3348/kjr.2017.18.4.570
Mak KK, Pichika MR (2019) Artificial intelligence in drug development: present status and future prospects. Drug Discov Today 24(3):773–780. https://doi.org/10.1016/j.drudis.2018.11.014
McCarthy A (2020) The biomarker future is digital. www.clinicalomics.com. Accessed 21 July 2020
Merk D, Friedrich L, Grisoni F et al (2018) De novo design of bioactive small molecules by artificial intelligence. Mol Inform 37(1–2):1700153. https://doi.org/10.1002/minf.201700153
Michelle (2019) The growth of artificial intelligence (AI) in healthcare. www.healthrecoverysolutions.com. Accessed 6 July 2020
Mohs RC, Greig NH (2017) Drug discovery and development: role of basic biological research. Alzheimer's & dementia: Translational Research & Clinical Interventions. Alzheimers Dement (N Y) 3(4):651–657. https://doi.org/10.1016/j.trci.2017.10.005
Nolting J (2006) Developing a neural network model for health care. AMIA Annu Symp Proc 2006:1049
Park JA, Chang WA (2001) Assessment of HIV/AIDS-related health performance using an artificial neural network. Information & Management. Inf Manag 38(4):231–238. https://doi.org/10.1016/S0378-7206(00)00068-9
Patel UK, Anwar A, Saleem S et al (2019) Artificial intelligence as an emerging technology in the current care of neurological disorders. J Neurol. https://doi.org/10.1007/s00415-019-09518-3
Patrício M, Pereira J, Crisóstomo J et al (2018) Using Resistin, glucose, age and BMI to predict the presence of breast cancer. BMC Cancer 18:29. https://doi.org/10.1186/s12885-017-3877-1
Pawar S, Liew TO, Stanam A et al (2020) Common cancer biomarkers of breast and ovarian types identified through artificial intelligence. Chem Biol Drug Des 96(3):995–1004. https://doi.org/10.1111/cbdd.13672
Pupale R (2018) Support vector machines (SVM) — an overview. https://towardsdatascience.com/https-medium-com-pupalerushikesh-svm-f4b42800e989. Accessed 18 July 2020
Raghavendra U, Acharya UR, Adeli H (2019) Artificial intelligence techniques for automated diagnosis of neurological disorders. Eur Neurol 82:41–64. https://doi.org/10.1159/000504292
Ray S (2017) Understanding support vector machine (SVM) algorithm from examples (along with code). https://www.analyticsvidhya.com/blog/2017/09/understaing-support-vector-machine-example-code/. Accessed 25 July 2020
Reddy V (2019) Using AI to identify biomarkers that facilitate personalized medicine. https://www.proxzar.ai/blog/using-ai-to-identify-biomarkers-that-facilitate-personalized-medicine/. Accessed 25 July 2020
Rehme AK, Volz LJ, Feis DL et al (2015) Identifying neuroimaging markers of motor disability in acute stroke by machine learning techniques. Cereb Cortex 25(9):3046–3056. https://doi.org/10.1093/cercor/bhu100
Schmitt (2020) Artificial intelligence in medicine. www.datarevenue.com. Accessed 18 July 2020
Schneider G (2017) Automating drug discovery. Nat Rev Drug Discov 17:97–113. https://doi.org/10.1038/nrd.2017.232
Srivastava T (2014) How does artificial neural network (ANN) algorithm work? Simplified! www.analyticsvidhya.com. Accessed 23 July 2020
The Medical Futurist (2018) What Do Digital Biomarkers Mean? https://medicalfuturist.com/what-do-digital-biomarkers-mean/. Accessed 26 July 2020

Tsoulos IG, Mitsi G, Stavrakoudis A et al (2019) Application of machine learning in a Parkinson's disease digital biomarker dataset using neural network construction (NNC) methodology discriminates patient motor status. Front ICT. https://doi.org/10.3389/fict.2019.00010

Vanneste S, Song JJ, De Ridder D (2018) Thalamocortical dysrhythmia detected by machine learning. Nat Commun 9:1103. https://doi.org/10.1038/s41467-018-02820-0

Vijay SS (2013) Applicability of artificial intelligence in different field of life. Int J Sci Eng Res 1 (1):28–35

Villar JR, Gonzalez S, Sedano J et al (2015) Improving human activity recognition and its application in early stroke diagnosis. Int J Neural Syst 25(4):1450036. https://doi.org/10.1142/S0129065714500361

West DM (2018) What is artificial intelligence? www.brookings.edu. Accessed 10 July 2020

Wu J (2019) AI, machine learning, deep learning explained simply. towardsdatascience.com. Accessed 4 July 2020

Zheng G, Patolsky F, Cui Y et al (2005) Multiplexed electrical detection of cancer markers with nanowire sensor arrays. Nat Biotechnol 23:1294–1301. https://doi.org/10.1038/nbt1138

Zhu T, Cao S, Su PC et al (2013) Hit identification and optimization in virtual screening: practical recommendations based on a critical literature analysis. J Med Chem 56(17):6560–6572. https://doi.org/10.1021/jm301916b

Automated Diagnosis of Diabetes Mellitus Based on Machine Learning

2

Abstract

According to the ninth edition of IDF Diabetes Atlas 2019, the worldwide prevalence of diabetes mellitus in 2019 was 463 million and has been estimated to escalate to 700 million, owing to a 51% increase in diabetes cases by 2045. It consequentially increases the risk of cardiovascular diseases by 3%, nephropathy by 5.9% and neuropathy by 10%. Another troublesome factor is the healthcare expenditure for diabetes management; the average diabetes-related health expenditure per person has multiplied 2.38-folds between the years 2010 and 2019. This chapter aims to enhance our understanding on the predictability of the onset of diabetes mellitus. We have developed built multiple (no.) machine learning models based on Pima Indians of Arizona, a niche which is highly susceptible to diabetes, and sourced the dataset from the National Institute of Diabetes and Digestive and Kidney Diseases, which will help the readers to understand that this emerging information technology is becoming society's most progressive tool and further may effectively use the information for their research endeavours.

Keywords

Diabetes mellitus · Artificial intelligence · Machine learning · Classification

2.1 Introduction

Diabetes mellitus is a chronic metabolic disorder that causes hyperglycaemia, resulting from impairments in insulin secretion, insulin action or both. Ironically, this metabolic disease is linked with long-term damage, dysfunction and failure of multiple organs, especially the eyes, kidney, nervous system, heart and blood vasculature. Globally 422 million people suffer with diabetes, and the majority of them belong to low- and middle-income countries and also attribute 1.6 million

A. Saxena, S. Chandra, *Artificial Intelligence and Machine Learning in Healthcare*,
https://doi.org/10.1007/978-981-16-0811-7_2

deaths every year. According to WHO projects, diabetes deaths will double between 2005 and 2030 (Sarwar et al. 2010).

Over the years, developments in information technology, statistics and computer has inspired many researchers to employ computational methods and multivariate statistical studies to analyse disease prognostics, which outpace the accuracy of empirical studies. This chapter will highlight the artificial intelligence (AI) approach especially machine learning for diabetic predictions. We explore how AI assists diabetic diagnosis and prognosis, specifically with regard to its unprecedented accuracy, which is even higher than that of general statistical application. We also constructed the model-based different approaches and attributes. Finally, comparison has been executed based on classification of model, which can be considered as clinical implementation of AI. Hence, this chapter delivers a novel perspective on how AI can expedite automated diabetic diagnosis and prognosis, contributing to the improvement of healthcare in the future.

2.2 Diabetes Mellitus

Diabetes mellitus is an incurable, metabolic disorder triggered by faulty insulin secretion and insulin resistance along with alterations of protein and lipid metabolism, which results in chronic hyperglycaemia. Prolonged hyperglycaemic conditions lead to glycation of proteins which subsequently leads to secondary pathological manifestations that affect the eyes, kidneys, nerves and arteries (Kharroubi and Darwish 2015). Glycation carried out by monosaccharides damages cells by impairing the function of target proteins, adds to oxidative stress and activates lethal signal transduction pathways (Taniguchi et al. 2015). Symptoms of diabetes mellitus include frequent urination, unexplained weight loss, excessive thirst, fatigue, numbness or tingling in the feet and hand tips and dry skin and sores (American Diabetes Association n.d.).

2.2.1 Classification of Diabetes Mellitus

Prediabetes is an intermediate state of hyperglycaemia with glycaemic parameters above normal but below the diabetes threshold. Characterization of the underlying pathophysiology is much more developed in type 1 diabetes mellitus than in type 2 diabetes mellitus (Mellitus 2006). The global disease burden of the major types of diabetes is shown in Fig. 2.1.

It is aetiologically classified into three categories:

1. *Type 1 diabetes mellitus* is an autoimmune disease contributing to approximately 5% of diabetic cases with a high prevalence in adolescents. It is majorly caused due to destruction of pancreatic islet cells via humoral response and T-cell-mediated inflammatory response. The presence of autoantibodies such as GAD65 glutamic acid decarboxylase, autoantibodies to insulin, IA2 and IA2β

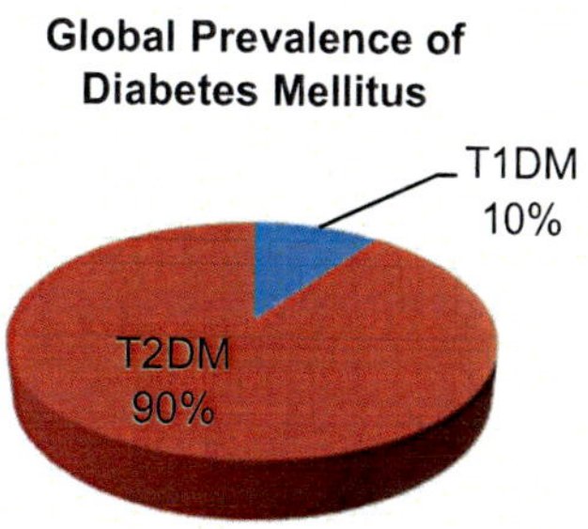

Fig. 2.1 Global prevalence of diabetes mellitus (Source: American Diabetes Association)

protein tyrosine phosphatase and ZnT8A zinc transporter protein against the pancreatic β cells is the diagnostic of this disease. These individuals are highly susceptible to ketoacidosis. Risk factors for type 1 diabetes are family history (genetic predisposition), autoantibodies, environmental factors, dietary deficiencies such as low vitamin D and specific geographic locations such as Sweden and Finland (Knip et al. 2005).

2. *Type 2 diabetes*, also termed as *non-insulin-dependent condition* that majorly affects the adults (20–79 yr. old), contributing to 95% of prevailing diabetes cases, is associated with obesity and insulin resistance that leads to decreased insulin production overtime. Study suggests that insulin resistance might improve with weight loss and treatment of hyperglycaemia but it can rarely be brought to normal levels. There are multiple risk factors for type 2 diabetes and prediabetic individuals: obesity, physical inactivity, age, susceptible race (Hispanics, American Indians), hypertension, polycystic ovarian syndrome, gestational diabetes mellitus and anomalous cholesterol and triglyceride levels.
3. *Gestational diabetes mellitus* is observed in 7% of all pregnant women worldwide. It often occurs due to hormonal imbalances that occur during pregnancy, leading to insulin resistance that subsides after pregnancy (American Diabetes Association 2015). Common risk factors for GDM are age, obesity, family history and susceptible race.

2.2.2 Diagnosis of Diabetes Mellitus

Diagnosis of diabetes mellitus is conducted on the basis of plasma glucose levels which comprises of either the fasting plasma glucose (FPG) and the 2-hr plasma glucose (2-hr PG) level during a 75-g oral glucose tolerance test (OGTT) or the A1C test. A random plasma glucose test is performed for individuals showing typical symptoms of hyperglycaemia (American Diabetes Association 2020). Table 2.1 summarises the levels of these diagnostic tests.

Table 2.1 List of pathological investigation for diabetes mellitus

S. no.	Test	Criteria
1.	FPG	≥126 mg/dL (7.0 mmol/L) *Fasting conditions referred here as zero calorie consumption for 8 h or more
2.	2-h PG during OGTT	≥200 mg/dL (11.1 mmol/L)
3.	A1C or HbA1c	≥6.5% (48 mmol/Mol)
4.	RPG	≥200 mg/dL (11.1 mmol/L)

2.2.3 Diabetes Management

The early diagnosis and management of diabetes mellitus is mandatory to prevent lethal complications associated with diabetes. Prolonged diabetes mellitus can lead to neuropathy, nephropathy, retinopathy, hearing impairments, skin infections and cardiovascular vascular diseases such as coronary artery disease, atherosclerosis and heart strokes (Health and Social Care Information Centre n.d.).

Pharmaceutical Therapy

The basic medication provided to individuals suffering from type 1 diabetes mellitus is insulin treatment to counter this autoimmune disease. In patients where this therapy is inefficacious, beta cell transplants and autoimmune blocking drugs are in clinical trials (Szadkowska et al. 2006; Szadkowska et al. 2008; Pietrzak et al. 2009). Metformin is the most commonly used medication, along with sodium-glucose co-transporter 2 inhibitors sold under the names phlorizin, dapagliflozin, amylin analogues, glucagon-like peptide 1 receptor agonists (exenatide, liraglutide) and di-peptidyl peptidase-4 (saxagliptin, vildagliptin) inhibitors (Frandsen et al. 2016; George and McCrimmon 2013; Otto-Buczkowska and Jainta 2018). Non-glycaemic treatments employ angiotensin-converting enzyme inhibitors, such as ramipril, often used in the cases of patients with nephropathy (National Collaborating Centre for Chronic Conditions (UK) 2008).

Self-Monitoring

Structured and personalised self-monitoring of blood glucose (SMBG) is an organised method of observing glucose levels that reveals glycaemic patterns throughout the day. Presently it is theorised that glycaemic variability contributes to diabetes complications independently of glycosylated haemoglobin (HbA1c) levels. To assess diurnal glucose excursions, SMBG has also been established as a useful tool as it helps to monitor diet control and treatment response and in general increases a patient's understanding of hypoglycaemia, thereby reducing their anxiety (J Meneses et al. 2015; Kirk and Stegner 2010; Schnell et al. 2013).

Although the pharmacological management of diabetes is sought after and provides several therapeutic opportunities, particularly in the type 2 diabetes mellitus, the changes in the lifestyle are essential: by maintaining proper diet and physical activities, one can reduce obesity associated with this type of diabetes (Nathan et al. 2009).

2.3 Role of Artificial Intelligence in Healthcare

Computational and artificial intelligence (AI) is an approach of enabling a computer system or software to think like an intelligent human being. Intelligence is defined as the capability of a system to perform tasks such as calculations, reasoning, perceiving relationships and analogies, learn from experiences, storing and retrieving data based on memory, solving problems, comprehending complex ideologies, employing natural language processes fluently and classifying, generalizing, and adapting to new states (Russell and Norvig 2002). Due to its diverse nature, AI is exploited by biologists across the globe to solve complex biological problems by applying algorithms to massive biological data obtained after experimental studies (Narayanan et al. 2002). From bioimaging, signal detection, sequencing analysis, protein structure folding to molecular modelling for drug discovery, artificial intelligence improves the practices of computational biology to yield cheaper yet accurate solutions (Nápoles et al. 2014).

AI-based technologies have proved to aid physicians to study complex diseases by parallel comparison of different cases of a disease through a single application/tool. Such applications assist in the detection and diagnosis of diseases within the early stages of progression, to come up with answers for complex cases in an easy, precise, quicker and overall accurate analysis to predict the future trends of a specific disease. This enables medical professionals to decide accurate surgical or diagnostic procedures by employing time and motion studies (Davenport and Kalakota 2019). Computational tools support optimization of factors by which the origin of a disease can be identified. The traditional methods for disease detection are time-consuming and expensive as they employ skilled experts and require continuous monitoring and observations. Image separation, feature extraction, classification and prediction of diseases can be efficiently done by employing machine learning approaches. After the early detection of diseases, these computational programs support in precise diagnostics of disease by providing virtual assistance, robotic surgery, POC, etc. to improve the time of diagnostics (Vashistha et al. 2018).

AI can be employed to improve clinical trials of novel stem cell and gene therapeutics in patients by detailed designing of treatment procedures, estimating clinical outcomes, streamlining enlistment and maintenance of patients, learning based on input data and applying to new data, thereby reducing their complexities and cost. Supplementing human intelligence with artificial intelligence will have an exponential influence on continual development in multiple fields of medicine (Ruff and Vertès 2020; Vamathevan et al. 2019).

AI plays a very important part in personalised medicine, drug discovery and development and gene editing therapies. It acts an interface between clinical image flow and archived image data, which does not need application-specific designing to utilise it. AI-based disease diagnostic systems expedite decision-making, reduce rate of false positives and therefore provide improved accuracy in the detection of diverse diseases (Ahmed et al. 2020).

2.4 AI Technologies Accelerate Progress in Medical Diagnosis

There are few successful examples of artificial intelligence-based disease diagnosis. Each of these machine learning studies employs different algorithm; however, the fundamental idea remains the same. Figure 2.2 describes a basic disease diagnostic AI model right from the curating a database patient's test reports to predicting diagnostic outcomes for every disease.

A. B. Suma designed a machine learning application, which offers a cost-effective, non-invasive and radiation less approach for the early diagnosis of rheumatoid arthritis based on thermography approach. The application compares multiple segmentation algorithms to identify the most appropriate segmentation algorithm for the input thermal image. Three different image segmentation algorithms were utilised to extract the hotspot area and subsequently compared to the original thermograph to determine the effective segmentation algorithm in the detection of RA. The accuracy acquired by the model was 93% (Langs et al. 2008).

Jhajharia et al. conducted prognosis model for breast cancer cases based on artificial neural network algorithm with principal component analysis of processed parameters. They employed a multivariate statistical technique along with the neural network to develop the prediction model (Jhajharia et al. 2016). Principal component analysis performs preprocessing and feature extraction of the input data in the most pertinent system for training model. The ANN learns the patterns within the dataset

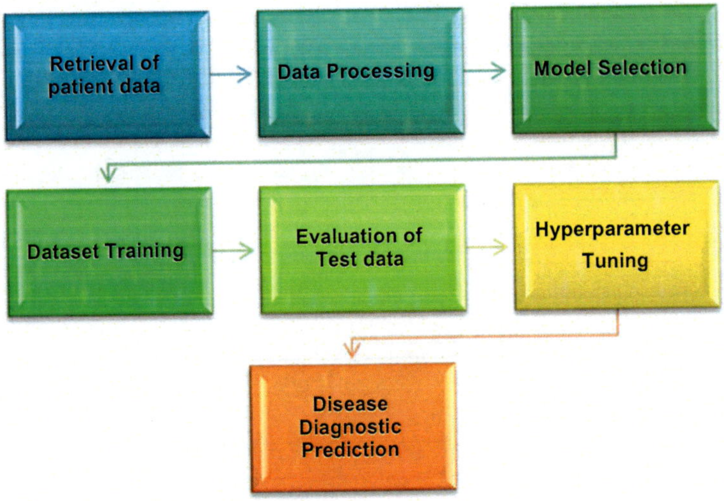

Fig. 2.2 Basic flow chart of a disease diagnostic AI model

for classification of new data. The accuracy from this ANN-based classification model was 96%.

Juan Wang developed a deep learning-based model for the detection of cardiovascular diseases. This model consists of a 12-layer convolutional neural network to distinguish breast arterial calcifications (BAC) from non-BAC and applies a pixel-wise, patch-based method for BAC identification. The performance of the system is evaluated by employing both free-response receiver operating characteristic (FROC) analysis and calcium mass estimation (Parthiban and Srivatsa 2012). The FROC analysis indicates that the deep learning technique achieved a level of detection comparable to the human experts. The calcium mass quantification test revealed that the inferred calcium mass is close to the actual values showing a linear regression, which yields a coefficient of determination of 96.24%.

Parthiban and Srivatsa (Challa et al. 2016) designed a machine learning model for the diagnosis of heart diseases. By using naive Bayes algorithm, an accuracy of 74% was achieved. SVM provided the highest accuracy of 94.60. K. N. Reddy and his co-worker have created an automated diagnosis model for Parkinson's disease by employing multilayer perceptron, random forest, Bayes network and boosted logistic regression (Burkov and Lutz 2019). Amongst the four models boosted logistic regression algorithm obtained the highest accuracy of 97.16% with an n area under the ROC curve of 98.9%.

2.5 Machine Learning

Learning basically refers to the process of acquiring a specific skill or knowledge during a study or experience. When a machine is capable of reproducing this basic act, it is termed as machine learning. It is an application of computer sciences, specifically a branch of artificial intelligence, which allows a computer system to learn a specific piece of data and develop itself from this study without the need of explicit programming (Bishop 2006). One can infer from this process that machine learning operates in two steps, namely the training phase and testing phase (Hastie et al. 2009). A model is defined with some parameters present in a data pool where the system learns the parameters based on their relationships and inherent properties in the training phase. This model is tested on a new dataset to predict the learnt outcomes (Alpaydin 2020). The ultimate goal of the model is to make generalised yet accurate predictions in the future or descriptive to gain knowledge from new and large datasets or both (de Ridder et al. 2013; Rao and Gudivada 2018).

2.5.1 Types of Machine Learning

Machine learning is broadly categorised into four groups: supervised, unsupervised, semi-supervised and reinforcement learning (Lee 2019).

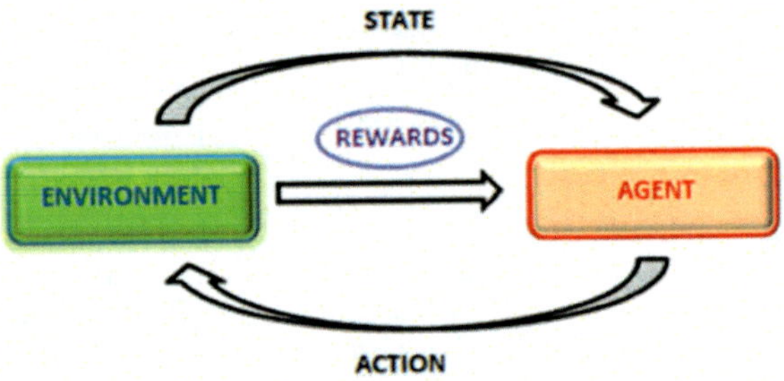

Fig. 2.3 Reinforcement learning architecture

Supervised Learning

- The dataset is a pool of labelled examples.
- The goal of a supervised learning is to use the dataset to produce a model that takes a feature vector x as input and output information that allows deducing the label for this feature vector.
- It majorly solves classification and regression problems.
- Decision trees, random forest, k-nearest neighbours and logistic regression are the examples of supervised machine learning algorithms.

Unsupervised Learning

- The dataset is a pool of unlabelled examples.
- The goal of an unsupervised learning is discover hidden pattern within the dataset where the output is not predefined.
- It can solve complex clustering and association problems.
- k-means for clustering and a priori algorithm for association are the examples of unsupervised machine learning algorithms.

Semi-Supervised Learning

- This combination will contain a very small amount of labelled data and a very large amount of unlabelled data.
- The goal of a semi-supervised learning algorithm is to improve supervised learning algorithm by using unlabelled data.
- It can solve problems of classification, regression, clustering and association.

Reinforcement Learning

- The machine is thriving in an environment where it recognises the state of that particular environment as feature vector in data.
- Each action brings different kind of rewards and can transfer the agent to another state (Sutton 1992).
- The goal of reinforcement learning is to make the system learn a policy.
- The policy is a function of the feature vector of a state that is considered as an input, and the outputs are an optimal action to implement in that state.
- If an action is ideal, it maximises the anticipated average reward. A simple figure describing the architecture of reinforcement learning is shown in Fig. 2.3 (Bishop 2006).

- Reinforcement machine learning resolves problems of sequential decision-making, where the goal is long term (Kesavadev et al. 2020).

2.5.2 Role of Machine Learning in Diabetes Mellitus Management

There are several applications of diabetes based on machine learning (Fig. 2.4).

Insulin Controller

An automated artificial pancreatic system improves the efficiency of glucose monitoring and liberates a patient from the hectic treatment regimen. Essentially the three major parts of an n artificial pancreas are the continuous glucose monitoring system, smart insulin controller and insulin delivery pumps (Bothe et al. 2013). For critical patients specifically studies are being conducted to develop algorithm for accurate insulin dose prediction and diet regimes that will be used as temporary management of glucose levels.

Reinforcement learning (RL) algorithms regulate insulin in a closed loop to deliver patient-specific insulin dosage plans that are responsive to the instant needs of the patients. RL provides advantage of expansion to infinite state sets, which allows the measurement of the variations in the glycaemic levels throughout CGM. However, the method has been vastly used in silico, so the success of RL algorithms for CGM in real patients (in vivo) is yet to be proved (Tyler et al. 2020). Tyler et al. have designed a k-nearest neighbours-based decision support system to detect causes of high and low glucose levels and offer weekly insulin dosage suggestions to T1DM patients taking multiple daily injection therapies (Donsa et al. 2015).

Lifestyle Support

Carbohydrate consumption and physical exercise are vital factors for managing diabetes mellitus. While the former raises the blood glucose values, the latter is

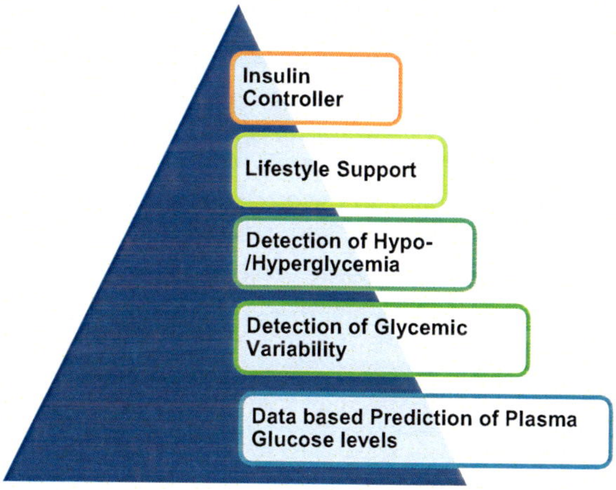

Fig. 2.4 Machine learning applications in diabetes management

glucose lowering (Anthimopoulos et al. 2014). In the era of Instagram and Facebook, clicking pictures of food has become a common practice. Anthimopolous designed GoCARB, an automated food-sensing mobile application for carbohydrate estimation in unpackaged foods, supporting T1DM patients. In this system the patient places a reference card next to their plate and captures two images of the same. These images are processed by linear SVC based on bag-of-features model, which reconstruct the 3D food item computationally. Finally, the quantity of food is estimated, and the amount of carbon, hydrogen and oxygen is calculated by merging the previous results and using the USDA nutrition database (Alfian et al. 2018).

Physical activity recognition is imperative for the estimation of energy expenditure. Alfian et al. proposed a bluetooth low energy-based sensor, which collects blood glucose, heart rate, blood pressure, weight and other personal data and stores this data in Apache Kafka, which undergoes real-time processing. Using this technology, one can observe existing body patterns and predict future changes in health based on multilayer perceptron classifier which is used to classify the diabetes patients metabolic rates; meanwhile, long short-term memory is used to estimate the blood glucose levels (Ellis et al. 2014). Ellis et al. developed a random forest classifier that predicts physical activity and energy consumed using accelerometers. In identification of physical activity, wrist devices performed better, whereas hip devices were well suited for energy consumption computation (Ghosh and Maka 2011).

Detection of Hypoglycaemia/Hyperglycaemia

The identification of hypoglycaemia and hyperglycaemia is considered as a characteristic classification problem. For a given set of input factors, the model should identify the occurrence of a hypoglycaemic or hyperglycaemic condition. The prediction can be condensed to a binary classification case, which is easier to predict than continuous predictions of blood glucose levels. Ghosh et al. propose a model based on the hybrid approach of non-linear autoregressive exogenous input modelling and genetic algorithm for deriving an index of insulin sensitivity (Seo et al. 2019). Machine learning can also be used to improve the accuracy of CGM systems. Seo et al. used machine learning algorithms (a random forest and support vector machine) using a linear function or a radial basis function, a k-nearest neighbour and a logistic regression to detect hypoglycaemia by utilising data-driven input factors (Qu et al. 2012).

Detection of Glycaemic Variability

Glycaemic variability is the fluctuations of blood glucose levels that indicate the quality of diabetes management due to increased risk of hypoglycaemic and hyperglycaemic episodes (Marling et al. 2013). Marling et al. employed a multilayer perceptron and support vector machine models for regressions on 250 CGM plots of 24 h on a consensus observed glycaemic variability metric, which has been manually classified into four CV classes (low, borderline, high or extremely high) by 12 doctors. The data underwent preprocessing by employing averaging and tenfold cross-validation prior evaluation. The support vector CPGV metric achieved an accuracy of 90.1%, with a sensitivity of 97.0% and a specificity of 74.1%, and outperformed other metrics such as MAGE or SD (Georga et al. 2011).

Data-Based Prediction of Plasma Glucose Levels

Data-based prediction of plasma glucose levels is categorised as a non-linear regression problem with input factors such as medications, dietary intake, physical activity, anxiety, etc. and blood glucose value as output value (Pappada et al. 2011). Pappada et al. showed a RMSE of 43.9 mg/dL in their study with ten type 1 diabetes mellitus patients using a neural network model. The model accurately identified 88.6% of normal glucose concentrations, 72.6% of hyperglycaemia but only 2.1% of hypoglycaemia correctly within a prediction range of 75 min. Many data-driven prediction methods lag behind in computation of hypoglycaemic and/or hyperglycaemic conditions because of the limited availability of data on hypoglycaemic and hyperglycaemic values (Dreiseitl and Ohno-Machado 2002).

2.6 Methodology for Development of an Application Based on ML

For predicting whether a patient is diabetic or not, there are five different algorithms: logistic regression, support vector machine, k-nearest neighbours, decision tree and random forest in machine learning predictive models, of which details are given in Fig. 2.5.

2.6.1 Dataset

The dataset used in this study has been originally obtained by the National Institute of Diabetes and Digestive and Kidney Diseases. The objective is to find whether a patient has diabetes or not, given certain values for different parameters. All the patients considered in this dataset are females above 21 years old. There are 768 instances available in this dataset. The independent parameters for this dataset are number of times the patient was pregnant, plasma glucose concentration level, diastolic blood pressure, triceps skinfold thickness, serum insulin in 2 h, body mass index, diabetes pedigree and age of the patient discussed in Table 2.2. There is a dependent variable outcome that tells if the patient is diabetic or not. Of these 768 instances, there are 268 instances of diabetes, and the rest of the instances are non-diabetic.

2.6.2 Data Preprocessing

The first step is to count the number of instances with missing values for each independent parameter. There are 227 missing values for the skin thickness parameter, which accounts for 30% of the total instances. Also there are 374 (49%) missing

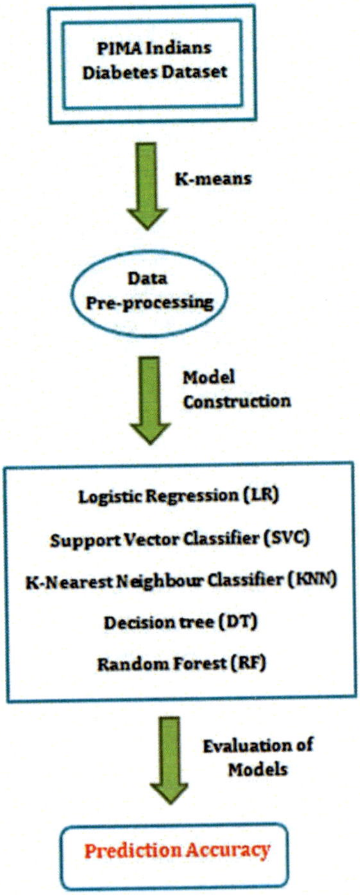

Fig. 2.5 Flow chart of methodology

Table 2.2 Attributes in Pima Indians dataset

S. no.	Attributes	Units	Type	Value range
1.	Pregnancy	No. of times pregnant	Integer	0–17
2.	Plasma glucose	mg/dL	Real	0–199
3.	Diastolic blood pressure	mmHg	Real	0–122
4.	Triceps skin fold	mm	Real	0–99
5.	Serum insulin	mu U/mL	Real	0–846
6.	Body mass index	kg/m^2	Real	0–67.1
7.	Diabetes pedigree		Real	0.078–2.42
8.	Age	Years	Integer	21–81

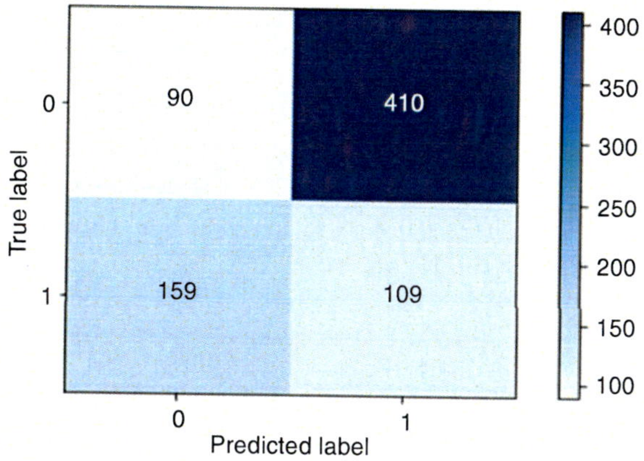

Fig. 2.6 Confusion matrix of k-means clustering

values for serum insulin in 2 h parameter. These two parameters are eliminated from our dataset as filling the missing values with placeholders could skew the classification model and decrease the accuracy of the model. For the rest of the parameters, missing values are replaced by substituting them with the median values.

For the model to give better accuracy, it could be helpful to look if there are any anomalies that could be weeded out before we train our model. k-means is a clustering algorithm that can be used for such purposes. With the help of this clustering algorithm, we can see if we can form two clusters and observe how well they can separate the instances into its respective prediction categories. The confusion matrix in Fig. 2.6 shows the misclassification after we apply k-means clustering to our 768 instances.

We can see that 569 instances were correctly clustered with a success rate of 74%. We keep these 74% instances, while eliminating the rest. This clears the anomalies in our dataset, and the classification model can give predictions with a greater confidence. The final step of the preprocessing involves standard scaling of all the values between 0 and 1 using min-max scaling.

2.6.3 Model Construction

Five different classification models have been created to see which model performs best. These classifiers are logistic regression (LR), support vector classifier (SVC), k-nearest neighbour (KNN) classifier, decision tree (DT) and random forest (RF). The parameters for all the models were declared such that the maximum accuracy could be acquired after tenfold cross-validation. For KNN, the best result was observed when the number of neighbours was set to 5. For RF, the maximum accuracy was obtained when the maximum depth was set to 4.

Logistic Regression

Logistic regression is a variant of linear regression. This model helps us to probabilistically model binary variables. This model is also called linear regression, which makes use of logit link. Logit here means the natural logarithm of an odds ratio. Logistic regression is quite useful when testing postulation of relationships between outcome dependent variables and one or more independent variables or parameters. The resultant plot while categorising instances of data appears linear in the middle but curved at the ends. This S-shaped plot is known as sigmoid. The advantages include faster computing due to low computational power requirements. Also we can make inference about relationships between independent parameters and output. The major disadvantage of logistic regression is that this models non-linear problem and often fails to capture complex relationships (Peng et al. 2002; Tambade et al. 2017).

Support Vector Machine

Support vector machine or SVM is a model for classification that can work well for linear and non-linear problems. To explain it in one line, the SVM algorithm creates an optimal hyperplane that separates the instances of data into different classes by building consistent estimators from data. Separate boundaries between instances of data are built by support vector machines by solving constrained quadratic optimization problems. Non-linearity and higher dimensions can be introduced in the model in different degrees with the help of various number of kernel functions available, also known as kernel trick. Most common kernels used when employing support vector machines are linear, rbf, poly and sigmoid. Generally learning algorithms works by learning characteristics that differentiate one classification from another. On the other hand, support vector machines find the most similar examples between classes also known as support vectors. Medical literature has reported that support vector machine models are on par or even exceed other machine learning algorithms (Nalepa and Kawulok 2019; Yu et al. 2020; Cristianini and Shawe-Taylor 2000; Schölkopf et al. 2002).

K-Nearest Neighbours

What differentiates k-nearest neighbours or KNN from other machine learning algorithms is that it directly uses instances of data for classification instead of first building a model. There is an adjustable parameter k that represents the number of nearest neighbours that are needed to estimate the membership of the class. No other information or details are required during the time of model construction. The estimate of class membership P(y|x) is the ratio of members of class y amongst the k nearest neighbours of x (Losing et al. 2016; Kotsiantis et al. 2007). Flexibility can be introduced with the help of altering the value of parameter k. Large values of k means less flexibility, while smaller number of k means more flexibility. The advantage of KNN is that the neighbours can explain the result after classification takes place. The disadvantage of KNN is that one can only define the parameter k with the help of trial and error as there is no other way to figure it out (Dasarathy 1991; Ripley 2007).

Decision Tree

In this algorithm, the instances of the dataset are split into treelike structures according to a set of criteria that results in maximization of separation of data. This tree or flow chart is made up of nodes that represent a test on an attribute, while each branch of the node represents the outcome of the test, and the leaf node represents the classification. This whole path from the root to individual leaf is said to make up the classification rules. Drawbacks include instability, i.e. a small change in data can significantly alter the structure of optimal decision trees. Also a multistep look ahead that considers different combinations of variables may result in different and sometimes even better classifications. The advantage is that this classification model is very easy to interpret since the classification rules are clearly defined by the flow chart (Breiman et al. 1984; Quinlan 1993).

Random Forest

This algorithm is an ensemble type of learning algorithm where it constructs multiple decision trees and outputs the class that is the mode of classes outputted by individual trees. Random forests tend to be better than decision trees since decision trees tend to overfit on training dataset while random forest algorithm provides a more generalised approach. It can also produce high-dimensional data by employing feature selection techniques. The disadvantages are that random forests are known to overfit on some noisy classification problems (Wyner et al. 2017; Biau et al. 2008).

2.6.4 Results

For the analysis of the performance of our models, tenfold cross-validation is done. This means that the instances were randomly divided into ten parts, where one part would be treated as the testing data, while the remaining nine parts would be treated as the training data. This process would be repeated ten times where each partition experiences a chance to be the testing data. The average of all the metrics such as accuracy, sensitivity, specificity and F1 score is taken to showcase the performance of our models. Support vector classifier, random forest, k-nearest neighbours, decision tree and logistic regression were the five models employed in this classification studies for the automated prediction of diabetes mellitus based on the Pima Indians dataset, and fortunately all five of them have displayed impressive accuracies

Table 2.3 Evaluation parameters of different predictive models

Classification model	Accuracy	Sensitivity	Specificity	F1 score
SVC	95.96	89.18	99.02	0.9361
RF	99.3	98.75	99.74	0.992
KNN	95.26	86.87	99.05	0.9236
DT	98.77	98.17	99.07	0.9857
LR	97.89	93.7	99.77	0.9659

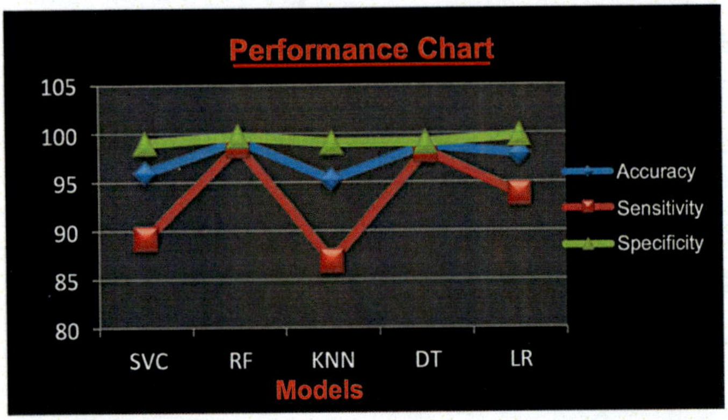

Fig. 2.7 Performance chart

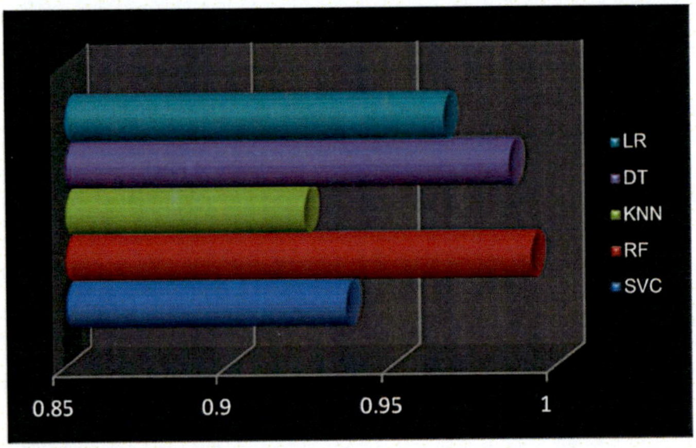

Fig. 2.8 F1 scores of the classification models

of prediction. Table 2.3 represents the evaluation parameters of the five models used in the study. The random forest model outperforms the other four models in terms of accuracy (99.3%), sensitivity (98.75%), specificity (99.74%) and F1 score (0.992) proves to be most suitable for the automated diagnosis of diabetes mellitus. A performance chart and F1 score distribution that compares all the five models based on their evaluation parameters are shown in Figs. 2.7 and 2.8.

2.7 Conclusion

Diabetes is a life-threatening metabolic disorder, which adversely affects the human body. Undiagnosed diabetes increases the risk of cardiovascular diseases, nephropathies and other chronic disorders. Therefore, the early detection of diabetes is vital for effective maintenance of a healthy life. Machine learning is a computational method for automated learning from experience and improves the performance to deliver better and accurate predictions. This provides an idea of recent artificial intelligent systems available for the detection and diagnosis of diabetes diseases. The system analyses the relevant medical imagery and associated point data to make an interpretation that can assist the physicians to make appropriate decisions in a clinical condition. The aim of this study was to make automated diabetes diagnosis available for everyone without the requirement of getting blood tests or visiting a hospital. The vision of this study was to provide accessible healthcare service promoting the idea of mHealth and affordable medical facilities. Also this automated study can be easily converted into a web-based application that one can easily access. However, it is to be noted that a web application is only a preliminary diagnosis. Any individual predicted to be at risk of diabetes must consult a certified physician to take proper tests and required medication immediately.

References

Ahmed Z, Mohamed K, Zeeshan S, Dong X (2020) Artificial intelligence with multi-functional machine learning platform development for better healthcare and precision medicine. Database 2020:baaa010

Alfian G, Syafrudin M, Ijaz MF, Syaekhoni MA, Fitriyani NL, Rhee J (2018) A personalized healthcare monitoring system for diabetic patients by utilizing BLE-based sensors and real-time data processing. Sensors 18(7):2183–3208

Alpaydin E (2020) Introduction to machine learning. MIT Press, Cambridge, MA

American Diabetes Association (2015) 2. Classification and diagnosis of diabetes. Diabetes Care 38 (Supplement 1):S8–S16

American Diabetes Association (2020) 2. Classification and diagnosis of diabetes: standards of medical Care in Diabetes-2020. Diabetes Care 43(Suppl 1):S14

American Diabetes Association (n.d.) Type 1 diabetes. https://www.diabetes.org/diabetes/type-1/symptoms

Anthimopoulos MM, Gianola L, Scarnato L, Diem P, Mougiakakou SG (2014) A food recognition system for diabetic patients based on an optimized bag-of-features model. IEEE J Biomed Health Inform 18(4):1261–1271

Biau G, Devroye L, Lugosi G (2008) Consistency of random forests and other averaging classifiers. J Mach Learn Res 9(Sep):2015–2033

Bishop CM (2006) Pattern recognition and machine learning. Springer, New York

Bothe MK, Dickens L, Reichel K, Tellmann A, Ellger B, Westphal M, Faisal AA (2013) The use of reinforcement learning algorithms to meet the challenges of an artificial pancreas. Expert Rev Med Devices 10(5):661–673

Breiman L, Friedman JH, Olshen RA, Stone CJ (1984) Classification and regression trees. Belmont, CA: Wadsworth. Int Group 432:151–166

Burkov A, Lutz M (2019) The hundred-page machine learning book. Notion Press, Chennai

Challa KNR, Pagolu VS, Panda G, Majhi B (Oct 2016) An improved approach for prediction of Parkinson's disease using machine learning techniques. In: 2016 international conference on signal processing, communication, power and embedded system (SCOPES). IEEE, Piscataway, NJ, pp 1446–1451

Cristianini N, Shawe-Taylor J (2000) An introduction to support vector machines and other kernel-based learning methods. Cambridge university press, Cambridge

Dasarathy BV (1991) Nearest neighbor (NN) norms: NN pattern classification techniques. IEEE Computer Society tutorial. IEEE Computer Society Press, Washington
Davenport T, Kalakota R (2019) The potential for artificial intelligence in healthcare. Future Healthc J 6(2):94–98
de Ridder D, de Ridder J, Reinders MJ (2013) Pattern recognition in bioinformatics. Brief Bioinform 14(5):633–647
Donsa K, Spat S, Beck P, Pieber TR, Holzinger A (2015) Towards personalization of diabetes therapy using computerized decision support and machine learning: some open problems and challenges. In: Smart health. Springer, Cham, pp 237–260
Dreiseitl S, Ohno-Machado L (2002) Logistic regression and artificial neural network classification models: a methodology review. J Biomed Inform 35(5–6):352–359
Ellis K, Kerr J, Godbole S, Lanckriet G, Wing D, Marshall S (2014) A random forest classifier for the prediction of energy expenditure and type of physical activity from wrist and hip accelerometers. Physiol Meas 35(11):2191
Frandsen CS, Dejgaard TF, Madsbad S (2016) Non-insulin drugs to treat hyperglycaemia in type 1 diabetes mellitus. Lancet Diabetes Endocrinol 4(9):766–780
Georga EI, Protopappas VC, Fotiadis DI (2011) Glucose prediction in type 1 and type 2 diabetic patients using data driven techniques. In: Knowledge-oriented applications in data mining. IntechOpen, London, pp 277–296
George P, McCrimmon RJ (2013) Potential role of non-insulin adjunct therapy in type 1 diabetes. Diabet Med 30(2):179–188
Ghosh S, Maka S (2011) Genetic algorithm based NARX model identification for evaluation of insulin sensitivity. Appl Soft Comput 11(1):221–226
Hastie T, Tibshirani R, Friedman J (2009) The elements of statistical learning: data mining, inference, and prediction. Springer, New York
Health and Social Care Information Centre (n.d.) National diabetes audit 2012–2013. Report 2: complications and mortality. http://www.hscic.gov.uk/catalogue/PUB16496/nati-diab-audi-12-13-rep2.pdf
J Meneses M, M Silva B, Sousa M, Sa R, F Oliveira P, G Alves M (2015) Antidiabetic drugs: mechanisms of action and potential outcomes on cellular metabolism. Curr Pharm Des 21 (25):3606–3620
Jhajharia S, Varshney HK, Verma S, Kumar R (Sept 2016) A neural network based breast cancer prognosis model with PCA processed features. In: 2016 international conference on advances in computing, communications and informatics (ICACCI). IEEE, Piscataway, NJ, pp 1896–1901
Kesavadev J, Saboo B, Krishna MB, Krishnan G (2020) Evolution of insulin delivery devices: from syringes, pens, and pumps to DIY artificial pancreas. Diabetes Ther 11(6):1251–1269
Kharroubi AT, Darwish HM (2015) Diabetes mellitus: the epidemic of the century. World J Diabetes 6(6):850–867
Kirk JK, Stegner J (2010) Self-monitoring of blood glucose: practical aspects. J Diabetes Sci Technol 4(2):435–439
Knip M, Veijola R, Virtanen SM, Hyöty H, Vaarala O, Åkerblom HK (2005) Environmental triggers and determinants of type 1 diabetes. Diabetes 54(suppl 2):S125–S136
Kotsiantis SB, Zaharakis I, Pintelas P (2007) Supervised machine learning: a review of classification techniques. Emerg Artif Intell Appl Comput Eng 160(1):3–24
Langs G, Peloschek P, Bischof H, Kainberger F (2008) Automatic quantification of joint space narrowing and erosions in rheumatoid arthritis. IEEE Trans Med Imaging 28(1):151–164
Lee D (17 Oct 2019) A brief introduction to reinforcement learning. https://medium.com/ai%C2%B3-theory-practice-business/reinforcement-learning-part-1-a-brief-introduction-a53a849771cf
Losing V, Hammer B, Wersing H (December 2016) KNN classifier with self adjusting memory for heterogeneous concept drift. In: 2016 IEEE 16th international conference on data mining (ICDM). IEEE, Piscataway, NJ, pp 291–300
Marling CR, Struble NW, Bunescu RC, Shubrook JH, Schwartz FL (2013) A consensus perceived glycemic variability metric. J Diabetes Sci Technol 7(4):871–879

Mellitus DIABETES (2006) Diagnosis and classification of diabetes mellitus. Diabetes Care 29: S43–S49
Nalepa J, Kawulok M (2019) Selecting training sets for support vector machines: a review. Artif Intell Rev 52(2):857–900
Nápoles G, Grau I, Bello R, Grau R (2014) Two-steps learning of fuzzy cognitive maps for prediction and knowledge discovery on the HIV-1 drug resistance. Expert Syst Appl 41 (3):821–830
Narayanan A, Keedwell EC, Olsson B (2002) Artificial intelligence techniques for bioinformatics. Appl Bioinforma 1:191–222
Nathan DM, Buse JB, Davidson MB, Ferrannini E, Holman RR, Sherwin R, Zinman B (2009) Medical management of hyperglycemia in type 2 diabetes: a consensus algorithm for the initiation and adjustment of therapy: a consensus statement of the American Diabetes Association and the European Association for the Study of diabetes. Diabetes Care 32(1):193–203
National Collaborating Centre for Chronic Conditions (UK) (2008) Type 2 diabetes: national clinical guideline for management in primary and secondary care (update). Royal College of Physicians, London
Otto-Buczkowska E, Jainta N (2018) Pharmacological treatment in diabetes mellitus type 1–insulin and what else? Int J Endocrinol Metab 16(1):e13008
Pappada SM, Cameron BD, Rosman PM, Bourey RE, Papadimos TJ, Olorunto W, Borst MJ (2011) Neural network-based real-time prediction of glucose in patients with insulin-dependent diabetes. Diabetes Technol Ther 13(2):135–141
Parthiban G, Srivatsa SK (2012) Applying machine learning methods in diagnosing heart disease for diabetic patients. Int J Appl Inform Syst 3(7):25–30
Peng CYJ, Lee KL, Ingersoll GM (2002) An introduction to logistic regression analysis and reporting. J Educ Res 96(1):3–14
Pietrzak I, Mianowska B, Gadzicka A, Mlynarski W, Szadkowska A (2009) Blood pressure in children and adolescents with type 1 diabetes mellitus–the influence of body mass index and fat mass. Pediatr Endocrinol Diabetes Metab 15(4):240–245
Qu Y, Jacober SJ, Zhang Q, Wolka LL, DeVries JH (2012) Rate of hypoglycemia in insulin-treated patients with type 2 diabetes can be predicted from glycemic variability data. Diabetes Technol Ther 14(11):1008–1012
Quinlan JR (1993) Program for machine learning. C4.5
Rao CR, Gudivada VN (2018) Computational analysis and understanding of natural languages: principles, methods and applications. Elsevier, Amsterdam
Ripley BD (2007) Pattern recognition and neural networks. Cambridge University Press, Cambridge
Ruff C, Vertès AA (2020) Harnessing in silico technologies to develop and augment second-generation cell-based therapies. In: Second generation cell and gene-based therapies. Academic Press, Cambridge, MA, pp 183–211
Russell S, Norvig P (2002) Artificial intelligence: a modern approach. Pearson, London
Sarwar N, Gao P, Seshasai SR, Gobin R, Kaptoge S, Di Angelantonio E, Ingelsson E, Lawlor DA, Selvin E, Stampfer M, Stehouwer CD (2010) Emerging risk factors collaboration diabetes mellitus, fasting blood glucose concentration, and risk of vascular disease: a collaborative meta-analysis of 102 prospective studies. Lancet 375(9733):2215–2222
Schnell O, Alawi H, Battelino T, Ceriello A, Diem P, Felton AM, Grzeszczak W, Harno K, Kempler P, Satman I, Vergès B (2013) Self-monitoring of blood glucose in type 2 diabetes: recent studies. J Diabetes Sci Technol 7(2):478–488
Schölkopf B, Smola AJ, Bach F (2002) Learning with kernels: support vector machines, regularization, optimization, and beyond. MIT press, Cambridge, MA
Seo W, Lee YB, Lee S, Jin SM, Park SM (2019) A machine-learning approach to predict postprandial hypoglycemia. BMC Med Inform Decis Mak 19(1):210–223
Sutton RS (1992) Introduction: the challenge of reinforcement learning. In: Reinforcement learning. Springer, Boston, MA, pp 1–3

Szadkowska A, Pietrzak I, Mianowska B, Markuszewski L, Bodalska-Lipińska J, Bodalski J (2006) Insulin resistance in type 1 diabetic children and adolescents--a simplified method of estimation. Endokrynologia, diabetologia i choroby przemiany materii wieku rozwojowego: organ Polskiego Towarzystwa Endokrynologow Dzieciecych 12(2):109–115

Szadkowska A, Pietrzak I, Mianowska B, Bodalska-Lipińska J, Keenan HA, Toporowska-Kowalska E, Młynarski W, Bodalski J (2008) Insulin sensitivity in type 1 diabetic children and adolescents. Diabet Med 25(3):282–288

Tambade S, Somvanshi M, Chavan P, Shinde S (2017) SVM based diabetic classification and hospital recommendation. Int J Comput Appl 167(1):40–43

Taniguchi N, Endo T, Hart GW, Seeberger PH, Wong CH (eds) (2015) Glycoscience: biology and medicine. Springer, Tokyo

Tyler NS, Mosquera-Lopez CM, Wilson LM, Dodier RH, Branigan DL, Gabo VB, Guillot FH, Hilts WW, El Youssef J, Castle JR, Jacobs PG (2020) An artificial intelligence decision support system for the management of type 1 diabetes. Nat Metab 2(7):612–619

Vamathevan J, Clark D, Czodrowski P, Dunham I, Ferran E, Lee G, Li B, Madabhushi A, Shah P, Spitzer M, Zhao S (2019) Applications of machine learning in drug discovery and development. Nat Rev Drug Discov 18(6):463–477

Vashistha R, Chhabra D, Shukla P (2018) Integrated artificial intelligence approaches for disease diagnostics. Indian J Microbiol 58(2):252–255

Wyner AJ, Olson M, Bleich J, Mease D (2017) Explaining the success of adaboost and random forests as interpolating classifiers. J Mach Learn Res 18(1):1558–1590

Yu D, Xu Z, Wang X (2020) Bibliometric analysis of support vector machines research trend: a case study in China. Int J Mach Learn Cybern 11(3):715–728

Artificial Intelligence in Personalized Medicine

3

Abstract

Personalized medicine is one of the largely considered approaches toward an accurate and safer treatment. At the same time, the medicine domain alone cannot maintain the modest outlay of the personalized medicine-centered treatment. Somehow, the accuracy of the medication and diagnosis using personalized medicine is lower when manualized than when involving artificial intelligence. Machine learning is one of the mostly used artificial intelligence models in convergence with high-throughput technologies. Natural language processing and robotics in convergence with machine learning are highly regarded in practicing an effective personalized medicine. Though machine learning is in the scenario of precision medicine first followed by personalized medicine, it still has to be accepted in the society for better development. This chapter gives an insight into how and where the artificial intelligence is used in the personalized medicine.

Keywords

Personalized medicine · Artificial intelligence · Medication · Diagnosis

3.1 Introduction

The field of medicine has grown significantly due to the integration of artificial intelligence (AI). Even the field of personalized medicine is being amalgamated with AI; however, it is still in its early stage and is facing a lot of issues (Awwalu et al. 2015). This fusion relies greatly on the algorithms of AI (Awwalu et al. 2015), but AI-driven platforms, AI-based analytics tools, etc. are also being used. For example, phenotypic personalized medicine (PPM) with the help of quadratic phenotypic optimization platform (QPOP) maximizes the desired outcome for combination

A. Saxena, S. Chandra, *Artificial Intelligence and Machine Learning in Healthcare*,
https://doi.org/10.1007/978-981-16-0811-7_3

therapy and their initial doses by selecting the drugs, and also PPM with the use of CURATE.AI dynamically recommends the most effective dosing approach: in the first case, QPOP is the AI-driven platform, whereas in the second case, CURATE.AI is the AI-driven platform (Blasiak et al. 2019). AI-based analytics tools are being extensively used to reduce the costs which arise while overcoming the huge amount of collected patient data, for example, Sapientia, Congenica's clinical genomic analysis platform, uses Exomiser (an AI-based analytics tool) to increase the speed of annotation and prioritization of variants from whole-exome sequencing (WES) in the diagnosis of rare diseases. Sapientia also enhances clinical decision-making by organizing the data in an easy manner, which helps to reduce the time taken for diagnosis by a huge margin (Suwinski et al. 2019). Other than the costs that arise while overcoming the huge amount of patient data, there are some other challenges that are faced when AI is being integrated with personalized medicine such as research costs, implementation costs, and government regulations. One of the major issues that is not being faced as of right now but may occur in the future is the threat of automation of the jobs of many healthcare personnel (Awwalu et al. 2015). Talking about the future, it may happen that by the use of AI-powered robotics one would be able to manufacture efficient and precise treatments, and it may also be possible that one would be able to foretell in which way the treatment strategy is going on the basis of AI-based simulation studies (Schork 2019). For the most part, it can be said that successful integration of AI in personalized medicine will save a lot of lives and may make the overall field of medicine impeccable.

3.2 Personalized Medicine

At present, most of our medicines follow a particular standard or a "one fits all" approach, despite the fact that various studies indicate that specific characteristics of an individual such as age, gender, height, weight, diet, and environment can influence the pharmacological effect of a drug. It has been found that even race plays a role in the responsiveness of a drug, for example, Blacks require higher concentrations of atropine and ephedrine to dilate their pupils when compared to the Mongols (Tripathi 2013).

The pharmacodynamics of a drug is also affected by the genetics of an individual, the dose of a drug needed to produce the same effect may vary by four- to sixfold among different individuals, and this is because of the differing rate of drug metabolism, which depends on the amount of microsomal enzymes present within the individual which again is genetically controlled (Tripathi 2013). As sometimes drugs have different effects on different individuals, it sometimes happens that a drug which provides the desired effect in one person causes an adverse effect or causes no effect at all in another person, even though it follows every standard. For example, a number of antihypertensive drugs interfere with the sexual function of men but not in women (Tripathi 2013). Another example can be of triflupromazine: its single dose induces muscular dystonias in some individuals but not in others (Tripathi 2013). Because of these reasons, healthcare practitioners are trying to find new ways to help their patients, and some of them are turning toward an emerging concept known as personalized medicine. It is referred to as "tailoring of medical treatment to the individualistic characteristics of each patient" (Tripathi 2013). It does not mean that a drug is created specifically for each patient, but rather it is a concept in which grouping of patients on the basis of their susceptibility toward a disease or their response toward a therapy is done (Tripathi 2013). This individualized approach helps the healthcare practitioners to provide their patients with a specialized treatment that is more precise, impactful, and efficient than the traditional treatment.

Personalized medicine is also sometimes referred to as stratified medicine or precision medicine. Precision medicine takes a group's common genetic patterns, their response toward drugs, their environment, and their lifestyles into account and provides the medical professionals with the information that they need to create specific treatments for their illnesses (Gameiro et al. 2018). Along with all of this, specific biomarkers are also taken into consideration: for example, Herceptin is used for the treatment of breast cancer when it is caused by the overexpression of HER-2 protein, whose biomarker is HER-2/neu receptor; Zelboraf is used for the treatment of melanoma when it is caused by defect in V600E, whose biomarker is BRAFV600E (Esplin et al. 2014).

Personalized dosing is also a very important part of personalized medicine as the standard adult dose is for medium-sized individuals. For children, unusually obese or lean individuals, the dose may be calculated on the basis of body mass index (Tripathi 2013). In one study, it was predicted that personalized dosing of warfarin could help in the prevention of 17,000 strokes and 43,000 emergency room visits in the USA; this prediction was later tested in 3600 patients, which resulted in 30% reduction in hospitalizations (Cutter and Liu 2012).

Another subset of personalized medicine that is newly emerging is "personalized sequencing"; it uses sequencing technologies such as whole genome sequencing and whole exome sequencing, and it also uses the data from the Human Genome Project. Personalized sequencing has advanced the way of studying and treating cancer. Some of the ways of personalized sequencing which have impacted cancer care are personalized tumor DNA sequencing, germline sequencing, and cancer cell DNA sequencing. An example of personalized tumor DNA sequencing impacting the treatment of cancer is the discovery of a loss-of-function mutation in *TSC1* in around 5% of advanced bladder cancer cases by the use of whole exome sequencing, and this was correlated with tumor sensitivity to everolimus, which suggested that these bladder cancer patients may be treated by everolimus therapy. As for germline sequencing, it helps to assess underlying patient risk which occurs due to known alterations and causes hereditary cancer predisposition syndromes such as Li-Fraumeni syndrome which further helps in the implementation of preventative measures and screening protocols for early detection. But, still, germline sequencing has not made much of an impression on cancer (Cutter and Liu 2012). Coming back to the vast area of personalized medicine, some other examples are the following:

- The dose of digestive enzyme supplement given during the treatment of cystic fibrosis is adjusted on the basis of volume and type of food ingested, number of meals, body mass gain, growth rate, type of enzyme used, and the response to the enzyme (Marson et al. 2017).
- For colorectal cancer patients with *KRAS* mutations, new treatments are being prepared as *KRAS* mutations are a predictive marker of resistance toward cetuximab and panitumumab (Pritchard and Grady 2011).
- By the use of gene expression profiling, acute myeloid leukemia patients are being grouped on the basis of their level of risk, according to which the intensity of their therapy is being tailored (Ken Redekop and Mladsi 2013).
- By the use of personalized topical therapeutics, it was found that the rate of healing of wounds had significantly increased both statistically and clinically (Dowd et al. 2011).
- Treatment of non-small cell lung cancer (NSCLC) patients with EGFR mutation with gefitinib led to longer progression-free survival, compared to NSCLC patients with no EGFR mutation (Jackson and Chester 2014).

Even though only a few personalized medicines are in practice, there is a need to incorporate this field into our clinical setting as it allows the patients to be treated with the most suitable medicines and therapies. This will lead to an improvement in the safety and efficacy of the drugs, as they will be tailored according to the needs of the subgroup, which will, in turn, lower the cases of adverse effects caused by drugs (Gurwitz and Manolopoulos 2018). Despite the fact that this field is new, it holds a lot of prospects; one of the reasons for this is the advancement in technology. For example, the development of diagnostic imaging for monitoring therapeutic efficacy can allow researchers and healthcare practitioners to select a therapy, plan a treatment, monitor an objective response, and plan a follow-up therapy, which will lead to the enhancement of the field of personalized medicine (Ryu et al. 2014). Another example of this can be theranostics, another emerging field, in which one pharmaceutical agent is used to diagnose disease, provide therapy, and monitor the progress of the treatment and the efficacy. This allows the monitoring of drug levels in targeted tissues and therapeutic response of the patient according to which the treatment can be adjusted to suit the needs of the patient, therefore leading to the concept of personalized medicine (Jo et al. 2016). Efforts are going on to personalize even the traditional Chinese medicines by the use of systems biology (Zhang et al. 2012). Also, a lot of work is going on in the development of personalized medicines for the treatment of chronic lymphocytic leukemia (Rozovski et al. 2014), smoking cessation (Nagalla and Bray 2016), thrombosis (Bierut et al. 2014), etc.

3.3 Importance of Artificial Intelligence

When a device is said to possess artificial intelligence (AI), it mimics the human intelligence. However, a hope that artificial intelligence can cede the human intelligence can make the future promising. Artificial intelligence can result in several outcomes such as reasoning, prediction, learning, and autocorrection. Artificial intelligence is widely used in every industry that needs the function of intelligence, but in healthcare, artificial intelligence has primarily became an extension to the traditional diagnosis; however, at present, it has surpassed the traditional way of diagnosis.

Though healthcare does not depend solely on AI, it is on the rise across the world. Due to the usage of AI in disease diagnosis, the early disease symptom prediction rate is already far from the average in the first world countries. The better treatment that is said to be given in the first and second world countries can be due to the use of AI in various aspects of diagnosis and treatment.

AI can be achieved by using different methods based on its functionality and recognition. Mostly used AI works on the basis of prediction and classification. The diverse nature of AI, which is an umbrella for different algorithms, statistical techniques, and learning models, assists not only the technical industry but also the biologists for better clinical manifestations.

The traditional approach of the medical diagnosis includes phenotype, morphological, and cytogenetic analysis. However, this conventional method is money and time exhaustive. As the biological world has taken a step toward accepting the branch of computer science, the conclusion of the diagnosis is often not answered in the early stages. Though AI has been in this world since 1950s, its evolution to be a part of mundane life took decades. Now, the role of AI in the healthcare is impressive as biologists can exploit the luxuries of predicting the early stages of a certain disease.

3.4 Use of Artificial Intelligence in Healthcare

With the evolving lifestyle, threat to the human life has been developing, and the medical world needs a leverage which can drive the diagnosis to attain maximum accuracy. The convergence of machine learning (ML) and different high-throughput technologies elevates the degree of diagnosis accuracy in the medical field. ML is a key method for higher disease prediction rate. Though ML can be effectively combined with many other predictive techniques, as shown in Fig. 3.1, ML is integrated hugely with other AI techniques such as robotics and vision in healthcare. ML and robotics have come together to surpass the conventional surgical procedures such as suturing. ML also helps the robot attain optimum workflow modeling by training the same. This training helps the robot increase the surgical skills and can effectively reduce the time spent on suturing.

Computer vision and machine learning together have improved recently. Image analysis and processing are two of the functions involved in computer vision. However, there is an interlude between healthcare and computer vision. This gap seems to be covered by introducing the machine learning algorithms into computer vision. The medical diagnoses in healthcare include image analysis for which computer vision can help thoroughly after being trained with the previous data. Prior the use of ML in medicine, high-throughput technologies such as

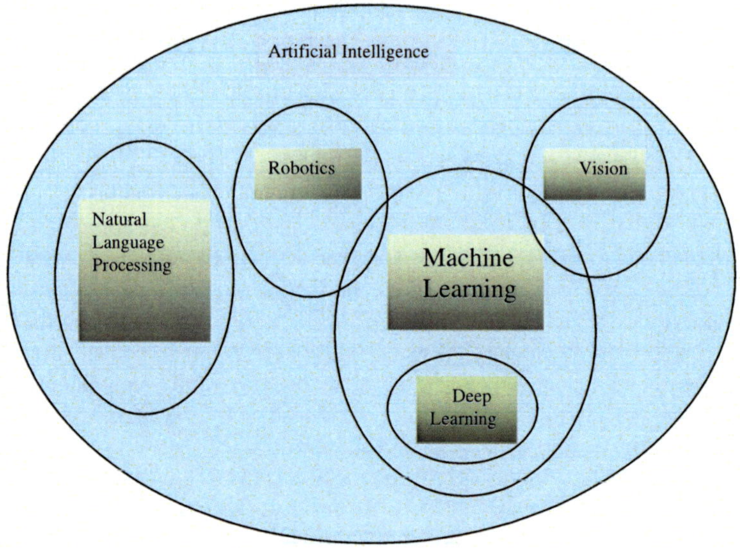

Fig. 3.1 Most commonly used models of artificial intelligence in healthcare

next-generation sequencing (NGS) which help in genotyping to detect chromosomal anomalies were elevated due to their rapid and cost-effective DNA/RNA sequencing. However, the use of ML in NGS has made the genotyping even more efficient and error free (Jiang et al. 2017).

Howsoever, the utilization of AI in the healthcare does not wipe out the conventional diagnoses or physicians. For ML to be in the picture, it needs data which is labeled/unlabeled. The input data, henceforth, has to be collected from clinical notes and medical diagnoses. But, certainly, ML can reduce the errors in the conventional techniques. One of the subgroups under the umbrella of AI includes natural language processing (NLP) as shown in Fig. 3.1. For a drug to be administered properly, it is important that the diagnoses done are accurate. Not only a drug cannot be administered solely on the basis of diagnoses but also the phenotypic and genetic factors of the patient. As the clinical record made by the physician is analyzed for the administration of drugs, the patient is often given "one fits all" drugs. Hence, the widely used approach in all the medically advanced countries is converting the clinical records to electronic medical reports (EMR). The EMRs are then subjected to the algorithms of ML for the prediction analysis on the basis of the genetic and phenotypic pattern of the patient besides the clinical records for accurate dosage of the drugs (Fernald et al. 2011; Borisov and Buzdin 2019).

3.5 Models of Artificial Intelligence Used in Personalized Medicine

ML and statistical genetics together can create wonders in the data-driven personalized medicine. ML is mainly staged into two phases as given below:

- Training phase.
- Predictive phase.

Training phase mainly involves in feeding the model with labeled/unlabeled data pool. The fed data is nothing but the prior clinical notes given by the physician or the traditional diagnoses and specific biomarkers for the disease. The device is trained by different algorithms in such a way that the parameters can be clustered or considered as individual to obtain a prediction. The predictive phase speculates the different possibilities of the outcomes on the basis of relationship between the inherent feature vector and the trained data pool by deducing a pattern or a relation implicitly.

As shown in Fig. 3.2, supervised learning (SL) is one of the mostly used AI models in the personalization of medicine so far. It mainly focuses on obtaining the outcome in one shot. The data set used for training is labeled or also known data. Most of the algorithms used in SL use regression analysis to obtain a linear complex combination between the feature vector and the trained parameters.

Though reinforcement learning (RL) is not as much used as SL, it is one of the exponential models in precision medicine. Unlike SL models, algorithms of RL are not exhaustive and are sequential in deducing the problems. They are worked using delayed feedback besides interacting with the environment by making behavioral decisions. Hence, they can be widely used in the automated medical care and diagnosis of an individual by personalizing the medicine.

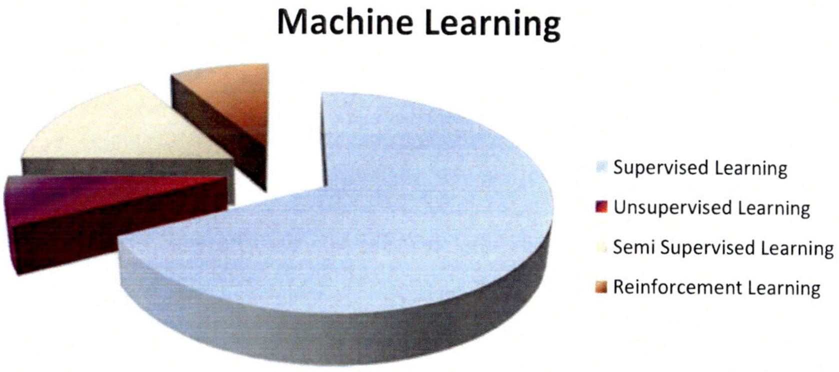

Fig. 3.2 Categories of machine learning used in personalized medicine. The data is obtained by the search of algorithms in PubMed

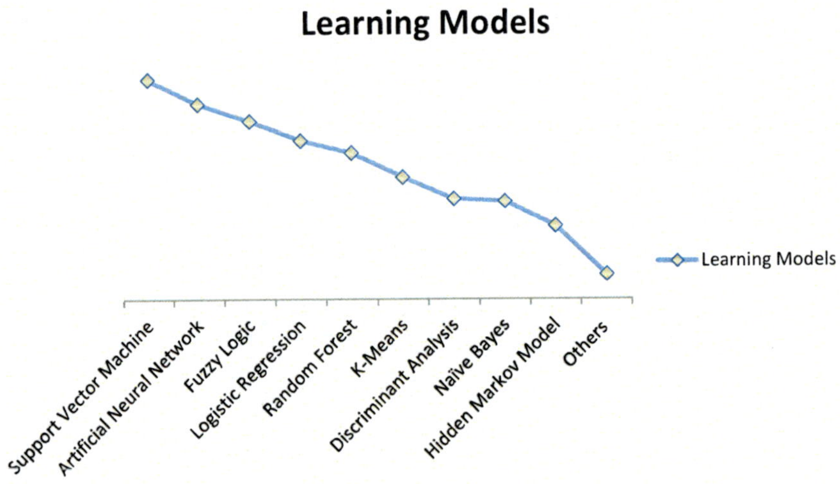

Fig. 3.3 Supervised and unsupervised learning models mostly used in personalized medicine. The data is obtained by the search of algorithms in PubMed

Unsupervised learning (USL) works different from SL in using unknown data sets. This model works with no previous experience and deduces the pattern and relationship of the parameters on its own. These algorithms can predict for more complex data than the SL ones. USL needs less manual labor than SL and can be used for better clustering. Personalized medicine needs more categorization of different factors such as biomarkers, microsomal enzymes, and lifestyle-based variables. This categorization can be eloquently done by USL algorithms. However, it is many times unpredictable than SL and RL, which is the reason it is not much used in personalizing the medicine (Wang et al. 2019).

Semi-supervised learning (SSL) is nothing but the algorithm that is trained with both known and unknown data sets. It is one of the mostly used learning models after SL. Precision medicine, unlike the personalized medicine, focuses only on the individual but not on the group. In progression of the disease prognosis, EMRs containing pedigree data along with cytogenetic data are some of the factors that are optimized by the SSL algorithms to obtain the degree of disease severity and origin of the disease in an individual (Fig. 3.3).

3.6 Use of Different Learning Models in Personalized Medicine

3.6.1 Naïve Bayes Model

This model comes under supervised learning. It is called naïve due to its use of strong independent comparison between the feature vector and input variables. Naïve Bayes model uses the Bayesian algorithm, which is developed in two phases as training phase and testing phase, respectively. This model performs multi-class predictions resulting in discriminant functions and probabilistic generative models.

Personalized medicine is in close relationship with pharmacogenomics when administration of the drug comes down to taking adverse drug reactions, molecular diagnostics, and classification of DNA into consideration. Hence, it is important to optimize every feature vector and consider these parameters individually to obtain better prediction models (Sampathkumar and Luo 2014).

It is known that thiopurine methyltransferase (TPMT) is a metabolic enzyme and participates in methylation of drugs such as azathioprine and 6-mercaptopurine that are widely used in treating autoimmune diseases. TPMT polymorphism results in adverse drug reactions due to the toxicological effects of the mentioned drugs. In such cases personalization of medicine comes into the picture. An individual screened with any of such anomalies cannot be put into the "one drug fits all" category. This is only one such example, but there are many syndromes which can lead the patient to death when the type of drug and its dosage are administered according to the standards and not on the basis of patient's independent factors (Katara and Kuntal 2016).

Bekir Karlik et al. developed a model using Bayesian algorithm for personalized cancer treatment. They used the pharmacogenetic data of TPMT polymorphism. They opted for naïve Bayes model as it can calculate probabilities of a single patient explicitly besides breaking the difficulty in "a priori" prediction. Their developed tool identified the TPMTs or SNPs for treating leukemia in the genome. They found utilization of naïve Bayes model more effective than the conventional DNA microarray to identify the polymorphisms that are responsible for adverse drug reactions (Karlık and Öztoprak 2012).

3.6.2 Support Vector Machine (SVM)

SVM is a part of supervised learning. SVM is being used in healthcare and mainly in personalized medicine since decades. SVM mainly involves in classification of the support vectors and thereby predicting the category of the new input data. The algorithms of SVM focus on regression analyses and categorization. SVM is highly advantageous in many cases as it does not only calculate the linear probability but also takes nonlinear data into consideration. SVM also is involved in fault or anomaly detection and hence is used widely in oncology (Grinberg et al. 2020).

The treatment of personalized oncology varies from the normal in few steps such as accurate prognosis according to the drug response. Breast cancer is one of the mostly affected cancers. However, breast cancer is not only due to the underlying etiology but can also be due to many different molecular etiologies that result in a malignant/benign tumor in the breast. Since many years, personalization of the treatment toward breast cancer has come into light for this very reason of having multiple subsets of molecular biomarkers that result in the disease. Millions of lives could have been saved by now if the biomarker targeted approach was used in the

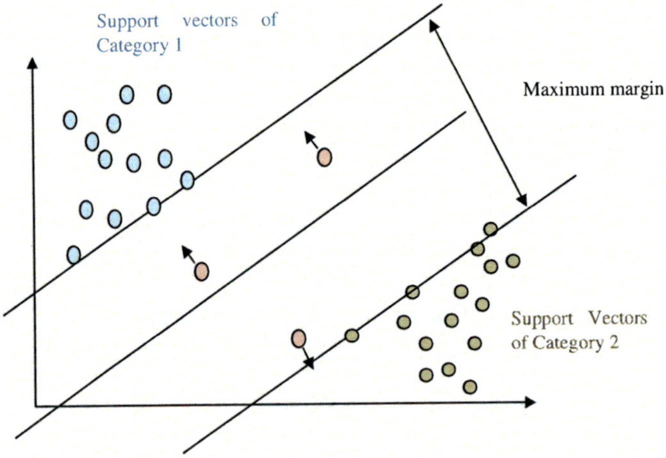

Fig. 3.4 Decision-making by classification in SVM

treatment. However, the conventional personalized medicine is expensive, and, therefore, it was unable to hit the ground running.

According to Mustafa Erhan Ozer et al., the use of SVM can accelerate personalized breast cancer treatment. They agreed to the point that the use of support vectors helps to classify high-dimensional big data effectively. One of the causes for breast cancer is the overexpression of HER-2 protein, for which the treatment must be targeted toward HER-2/neu receptor rather than the patient receiving a generalized treatment. Different breast cancer-causing factors which are considered support vectors in SVM are deduced from omics (transcriptomics, radiomics, genomics, proteomics) along with epidemiological data. When the problem is introduced, the algorithm classifies it into one of the categories as shown in Fig. 3.4.

3.6.3 Deep Learning

One of the mostly used deep learning techniques in personalized medicine is artificial neural network (ANN). The learning of these networks can be supervised, unsupervised, or semi-supervised. There are algorithms that are continuous and also discrete in ANN. Hence, ANN can perform not only classification but also clustering. Neural networks mimic the human neuron connections and are similarly not sequential unlike the regression models.

In personalized medicine, an individual's genotype or enzymology is considered. There might be many incidences where one or more parameters/problem data points were never labeled. Such cases cannot be accurately answered by supervised learning models, and, thereby, unsupervised learning has to be in the play.

The supervised learning of ANN needs large data sets, but they can self-extract and classify the features unlike other ML algorithms which need manual feature

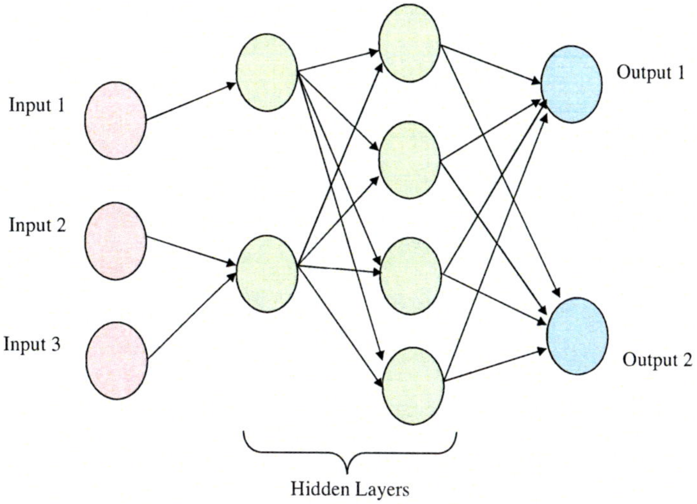

Fig. 3.5 Process of the ANN

extraction. ANN is a feed-forward network where input can be given only in the forward direction. ANN can be a single perceptron/layer or multiple perceptrons as shown in Fig. 3.5. It has one input layer where the data is input, single/multiple hidden layers where the data is processed, and an output layer which results in the decision. Linear/nonlinear properties in ANN are aggregated and weighed through the hidden layers, and, hence, any complex relationship can be found out effectively as shown in Fig. 3.5 (Papadakis et al. 2019).

However, ANN is poor in finding the gradient, and, hence, recurrent neural network (RNN) and convolution neural network (CNN) come into the picture. RNN and CNN propagate backward. The looping connection of weighing the data across the hidden perceptrons increases the accuracy of the output.

ANN can be used in optimization of the treatment, disease relapse prediction, accurate diagnosis, and many such other applications. Several researches show that cancer has been diagnosed accurately using the feature data. Few years ago, Microsoft has come up with the idea to diagnose and optimize the treatment using AI.

Naushad et al. developed an ANN model to predict breast cancer. They considered not only genetic polymorphisms but also nutrient and population-based variables into consideration. As discussed, the causes for breast cancer are many, and, hence, the biomarkers can be of different types. They investigated the susceptibility toward the cancer due to micronutrient modulation. The accuracy rate of this model came out to be 94.2% (Naushad et al. 2016).

Many other studies showed that when method combining ANNs in genetic algorithm, the results were very accurate and rapid. Personalized medicine heavily deals with sequencing one's DNA to obtain any anomalies or polymorphisms. The polymorphisms if any present in an individual would mostly lead to developing a disease, and, hence, any such have to be identified for early diagnosis. Many of the cancers include different molecular polymorphisms. Diagnosis is followed by treatment optimization, which can be accurately designed by the neural networks from weighing every parameter through different hidden nodes. ANNs are a step above the models that use linear regression as there is not much statistical knowledge that needs to be known beforehand while dealing with neural networks.

References

Awwalu J, Garba AG, Ghazvini A, Atuah R (2015) Application of Ai and soft computing in healthcare: a review and speculation. Int J Comput Theor Eng 7(6):439

Bierut LJ, Johnson EO, Saccone NL (2014) A glimpse into the future – personalized medicine for smoking cessation. Neuropharmacology 76:592–599. https://doi.org/10.1016/j.neuropharm.2013.09.009

Blasiak A, Khong J, Kee T (2019) CURATE.AI: optimizing personalized medicine with artificial intelligence. SLAS Technol 25:95–105. https://doi.org/10.1177/2472630319890316

Borisov N, Buzdin A (2019) New paradigm of machine learning (ML) in personalized oncology: data trimming for squeezing more biomarkers from clinical datasets. Front Oncol 9:658. https://doi.org/10.3389/fonc.2019.00658

Cutter GR, Liu Y (2012) Personalized medicine: the return of the house call? Neurol Clin Pract 2 (4):343–351. https://doi.org/10.1212/CPJ.0b013e318278c328

Dowd SE, Wolcott RD, Kennedy J, Jones C, Cox SB (2011) Molecular diagnostics and personalised medicine in wound care: assessment of outcomes. J Wound Care 20(5):232–239

Esplin ED, Oei L, Snyder MP (2014) Personalized sequencing and the future of medicine: discovery, diagnosis and defeat of disease. Pharmacogenomics 15(14):1771–1790. https://doi.org/10.2217/pgs.14.117

Fernald GH, Capriotti E, Daneshjou R, Karczewski KJ, Altman RB (2011) Bioinformatics challenges for personalized medicine. Bioinformatics (Oxford, England) 27(13):1741–1748. https://doi.org/10.1093/bioinformatics/btr295

Gameiro GR, Sinkunas V, Liguori GR, Auler-Júnior J (2018) Precision medicine: changing the way we think about healthcare. Clinics (Sao Paulo) 73:e723. https://doi.org/10.6061/clinics/2017/e723

Grinberg NF, Orhobor OI, King RD (2020) An evaluation of machine-learning for predicting phenotype: studies in yeast, rice, and wheat. Mach Learn 109:251–277. https://doi.org/10.1007/s10994-019-05848-5

Gurwitz D, Manolopoulos VG (2018) Personalized medicine. In: Reference module in chemistry, molecular sciences and chemical engineering. Elsevier, Oxford

Jackson SE, Chester JD (2014) Personalised cancer medicine. Int J Cancer 137(2):262–266. https://doi.org/10.1002/ijc.28940

Jiang F, Jiang Y, Zhi H et al (2017) Artificial intelligence in healthcare: past, present and future. Stroke Vascular Neurol 2:e000101. https://doi.org/10.1136/svn-2017-000101

Jo SD, Ku SH, Won Y-Y, Kim SH, Kwon IC (2016) Targeted Nanotheranostics for future personalized medicine: recent Progress in cancer therapy. Theranostics 6(9):1362–1377

Karlık B, Öztoprak E (2012) Personalized cancer treatment by using naive Bayes classifier. Int J Mach Learn Comput 2(3):339–344

Katara P, Kuntal HTPMT (2016) Polymorphism: when shield becomes weakness. Interdiscip Sci 8 (2):150–155. https://doi.org/10.1007/s12539-015-0111-1

Ken Redekop W, Mladsi D (2013) The faces of personalized medicine: a framework for understanding its meaning and scope. Value Health 16(6):4–9. https://doi.org/10.1016/j.jval.2013.06.005

Marson FAL, Bertuzzo CS, Ribeiro JD (2017) Personalized or precision medicine? The example of cystic fibrosis. Front Pharmacol 8:390. https://doi.org/10.3389/fphar.2017.00390

Nagalla S, Bray PF (2016) Personalized medicine in thrombosis: back to the future. Blood 127 (22):2665–2671. https://doi.org/10.1182/blood-2015-11-634832

Naushad SM, Janaki Ramaiah M, Pavithrakumari M, Jayapriya J, Hussain T, Alrokayan SA, Gottumukkala SR, Digumarti R, Kutala VK (2016) Artificial neural network-based exploration of gene-nutrient interactions in folate and xenobiotic metabolic pathways that modulate susceptibility to breast cancer. Gene 580(2):159–168. https://doi.org/10.1016/j.gene.2016.01.023

Papadakis GZ, Karantanas AH, Tsiknakis M, Tsatsakis A, Spandidos DA, Marias K (2019) Deep learning opens new horizons in personalized medicine (review). Biomed Rep 10:215–217. https://doi.org/10.3892/br.2019.1199

Pritchard CC, Grady WM (2011) Colorectal cancer molecular biology moves into clinical practice. Gut 60(1):116–129. https://doi.org/10.1136/gut.2009.206250

Rozovski U, Hazan-Halevy I, Keating MJ, Estrov Z (2014) Personalized medicine in CLL: current status and future perspectives. Cancer Lett 352(1):4–14

Ryu JH, Lee S, Son S, Kim SH, Leary JF, Choi K, Kwon IC (2014) Theranostic nanoparticles for future personalized medicine. J Control Release 190:477–484

Sampathkumar H, Chen XW, Luo B (2014) Mining adverse drug reactions from online healthcare forums using hidden Markov model. BMC Med Inform Decis Mak 14:91. https://doi.org/10.1186/1472-6947-14-91

Schork NJ (2019) Artificial intelligence and personalized medicine. Cancer Treat Res 178:265–283. https://doi.org/10.1007/978-3-030-16391-4_11

Suwinski P, Ong C, Ling MHT, Poh YM, Khan AM, Ong HS (2019) Advancing personalized medicine through the application of whole exome sequencing and big data analytics. Front Genet 10:49. https://doi.org/10.3389/fgene.2019.00049

Tripathi KD (2013) Essentials of medical pharmacology. JP Medical, New Delhi

Wang F, Wang Q, Nie F, Li Z, Yu W, Wang R (2019) Unsupervised linear discriminant analysis for jointly clustering and subspace learning. IEEE Trans Knowl Data Eng 3:3. https://doi.org/10.1109/TKDE.2019.2939524

Zhang A, Sun H, Wang P, Han Y, Wang X (2012) Future perspectives of personalized medicine in traditional Chinese medicine: a systems biology approach. Complement Ther Med 20 (1–2):93–99

Artificial Intelligence in Precision Medicine: A Perspective in Biomarker and Drug Discovery

4

Abstract

Clinical care is gradually transiting from the standard approach of "signs and symptoms" toward a more targeted approach that considerably trusts biomedical data and the gained knowledge. The uniqueness of this concept is implied by "precision medicine," which amalgamates contemporary computational methodologies such as artificial intelligence and big data analytics for achieving an informed decision, considering variability in patient's clinical, omics, lifestyle, and environmental data. In precision medicine, artificial intelligence is being comprehensively used to design and enhance diagnosis pathway(s), therapeutic intervention(s), and prognosis. This has led to a rational achievement for the identification of risk factors for complex diseases such as cancer, by gauging variability in genes and their function in an environment. It is as well being used for the discovery of biomarkers, that can be applied for patient stratification based on probable disease risk, prognosis, and/or response to treatment. The advanced computational expertise using artificial intelligence for biological data analysis is also being used to speed up the drug discovery process of precision medicine. In this chapter, we discuss the role and challenges of artificial intelligence in the advancement of precision medicine, accompanied by case studies in biomarker and drug discovery processes.

Keywords

Artificial intelligence · Biomarker · Diagnosis · Drug discovery · Omics data · Precision medicine · Prognosis

A. Saxena, S. Chandra, *Artificial Intelligence and Machine Learning in Healthcare*,
https://doi.org/10.1007/978-981-16-0811-7_4

4.1 Precision Medicine as a Process: A New Approach for Healthcare

Technological advancements facilitating advancement of omics-based diagnostics and therapeutics have the potential of creating the unprecedented ability for detection, prevention, treatment planning, and monitoring of diseases. The advent of modern computing (e.g., big data analytics, supercomputing, etc.) and new technological interventions (e.g., electronic health records, next-generation sequencing, etc.) is leading to the next generation of medicine and, in conjunction, delivering new tools for diagnostics, prognosis, and related clinical care (Pacanowski and Huang 2016). Historically, clinical care providers have continuously strived to provide better patient care in comparison to preceding generations by experimenting with the treatment procedures, bringing in innovative interventions, and gaining novel insights from clinical observations. Besides being a tedious process, the eventual goal was to provide a preemptive and precise treatment, which is beneficial for every patient. However, the availability of the multidimensional omics datasets along with the clinical data and evolving computational methodologies is achieving progressively more feasible patient care facilities considering individual patient's characteristics (Weil 2018). Consequently, the era of precision medicine, "an emerging approach for disease treatment and prevention that takes into account individual variability in genes, environment, and lifestyle for each person," is imminent (Burki 2017; König et al. 2017; Weil 2018).

The terms "precision medicine" and "personalized medicine" have been used synonymously, as there is supposed to be a lot of overlap between them. However, the National Research Council (USA) described preference of using the term "precision medicine" over "personalized medicine," as "personalized" could be misunderstood and suggest that treatments being developed are uniquely for each patient (Guide and Conditions 2015). Connoisseurs believed that clinical care providers have always been treating patients at a personalized level, taking into account factors such as age, gender, patient preferences, mobility levels, community resources, preexisting conditions, and other mitigating circumstances. In fact the personal approach has always been a part of a clinician-patient relationship and, therefore, cannot be considered as a completely new intervention, although it is an important and vital aspect of "precision medicine" (König et al. 2017). Its standard definition specifies that the treatment and diagnosis of a patient goes beyond the classical approach. However, the difference between the traditional method and true precision medicine is the availability and, most importantly, the degree of reliance on clinical data, lifestyle data, and especially genetic data and further biomarker information, which adds to this new approach of clinical care (Sankar and Parker 2017; Joyner and Paneth 2019). This approach of individually tailored healthcare provision on the basis of individual patient information is not new, as transfusion patients have been matched with donors according to blood type for more than a century, but currently growing availability of quality health data of all types has increased the chances manifold to make precise medicine a clinical reality.

The concept of precision medicine eliminates the "one size fits all" approach and strives giving patient cohorts treatment regimens, which are beneficial and with minimal/no side effects. Besides the genetic and clinical factors, the environmental features (the immediate physical surroundings, diet, lifestyle, etc.) also influence our health. With the combination of multidimensional and heterogeneous datasets, the knowledge gained may aid in potent treatment as well as planning for effective prevention and screening. Thus, precision medicine entails insight how elements from the environment interact with the genome, causing influencing variations and mapping the genotype-phenotype relationships. Imperatively its focus is not on the creation of person-specific drugs or medical devices but rather on the ability to classify individuals into cohorts or subpopulations that differ in their susceptibility for a particular disease or in their response to a specific treatment. Therefore, it needs to be emphasized that "precision" in "precision medicine" is being used in a colloquial sense, to mean both "accurate" and "precise" and not to be misinterpreted as implying unique treatments designed for each patient (Guide and Conditions 2015).

In the past 5 years, precision medicine has enabled key developments for complex diseases such as cancer, with the perspective of better understanding and facilitating predictive diagnosis as well as advancing prognosis. Availability of genetic tests and advanced diagnostics can indicate prospective therapeutic agents for distinct neoplasms in different tissues (Wang and Wang 2017). For example, in oncology, the detection of HER-2 indicating the treatment of breast cancer with trastuzumab is one of the most successful examples of precision medicine marker (Pinto et al. 2013), also the presence of the BCR/ABL or PML/RARA translocation, indicating specific treatments for leukemias, or the presence of V600E mutation, indicating specific treatment in melanomas (Deng and Nakamura 2017).

Pharmacogenomics has established drugs used in the treatment of infectious diseases which may show diverse consequences for the reason that genetic profiles differ in patients. This pharmacogenomic application indicated that because of this a few medications may cause adverse side effects and dosages need to be adjusted or the drug should be avoided for certain patients. In the treatment of a few viral diseases, detection of specific mutations causing resistance to antivirals has been recognized (Hauser et al. 2017). To determine the best treatment, presence of polymorphisms in genes involved in drug metabolism or in the major histocompatibility complex is important. Therefore, in addition to infectious diseases and cancer, researchers are also targeting metabolic diseases, for example, being optimistic to developing genetic tests to access and predict the risk of diseases such as type 2 diabetes and cerebrovascular diseases (Della-Morte et al. 2016; Scheen 2016; König et al. 2017).

Precision medicine is a complex process, involving numerous technologies to guide tailor-made patient diagnosis, prognosis, and treatment pathways. This is primarily reliant on distinctive data inputs such as clinical, genomic, lifestyle, and environmental features. Therefore, there is an imperative need of approaches for integrating, exploring, and translating the knowledge from these massive datasets diversely generated from the advancement of sequencing and other clinical technologies. Traditional approaches such as statistical analysis are helpful for such purposes; however, the use of artificial intelligence (AI) might be particularly appropriate for this setup. Further, with the evolution of high-performance computer capabilities, AI algorithms can achieve reasonable success, such as in predicting disease risk from the multidimensional and heterogeneous genomic and clinical datasets. AI applications with the focus on genomics, biomarker discovery (for patient diagnosis, prognosis, treatment pathway), and drug discovery are gradually leading in three major directions: generation of massive datasets with advanced analytics for novel insights, translating these insights into patient's bedside care, and edifice precision medicine. In this chapter, we review and discuss, in particular, how artificial intelligence has been used for biomarker and drug discovery, empowering precision medicine in emerging as a more precise and most suitable healthcare practice.

4.2 Role of Artificial Intelligence: Biomarker Discovery for Precision Medicine

The definition and use of biomarker have evolved over the years and may best illustrated as "characteristic that is objectively measured and evaluated as an indicator of normal biologic processes, pathogenic processes, or pharmacologic responses to a therapeutic intervention" (Atkinson et al. 2001; Slikker 2018). In clinical settings, the use of biomarkers have primarily wedged varied aspects associated to diagnosis and prognosis of diseases. The discovery of novel biomarkers provides a strategic opportunity for the advancement of healthcare and in reduction of associated costs. Therefore, they can be considered to play a key role in the development of precision medicine, providing a strategic opportunity for technological developments to improve clinical care (Slikker 2018). Considering its importance and to reduce obstacle in their development, US NIH (National Institutes of Health) and FDA (Food and Drug Administration) have developed the BEST (Biomarkers, EndpointS, and other Tools) resource, giving glossary of important definitions and their hierarchical relationships and capturing differences between biomarkers and their clinical assessments (Biomarker Working Group 2016; FDA-NIH Biomarker Working Group 2016). Artificial intelligence under these settings exploring the multidimensional datasets can thus accelerate the biomarker discovery process and provide a strategic opportunity for biomarker-driven therapeutic strategies, to improve human health and reduce the healthcare cost (Fig. 4.1).

Consequent to a particular disease, artificial intelligence can help inferring insights from the poly-omics (genome, epigenome, transcriptome, proteome, metabolome, microbiome) datasets in association with clinical and environmental factors. The prior knowledge with novel insights shall aid insinuating interactions and/or discovering relationship acumen into pleiotropy, complex interactions, and context-specific behavior. These multidimensional datasets can be trained using AI algorithms to discover relevant genotypic structures, which could be consequently mapped with a significant phenotype. Thereafter, it may be used for diagnostic

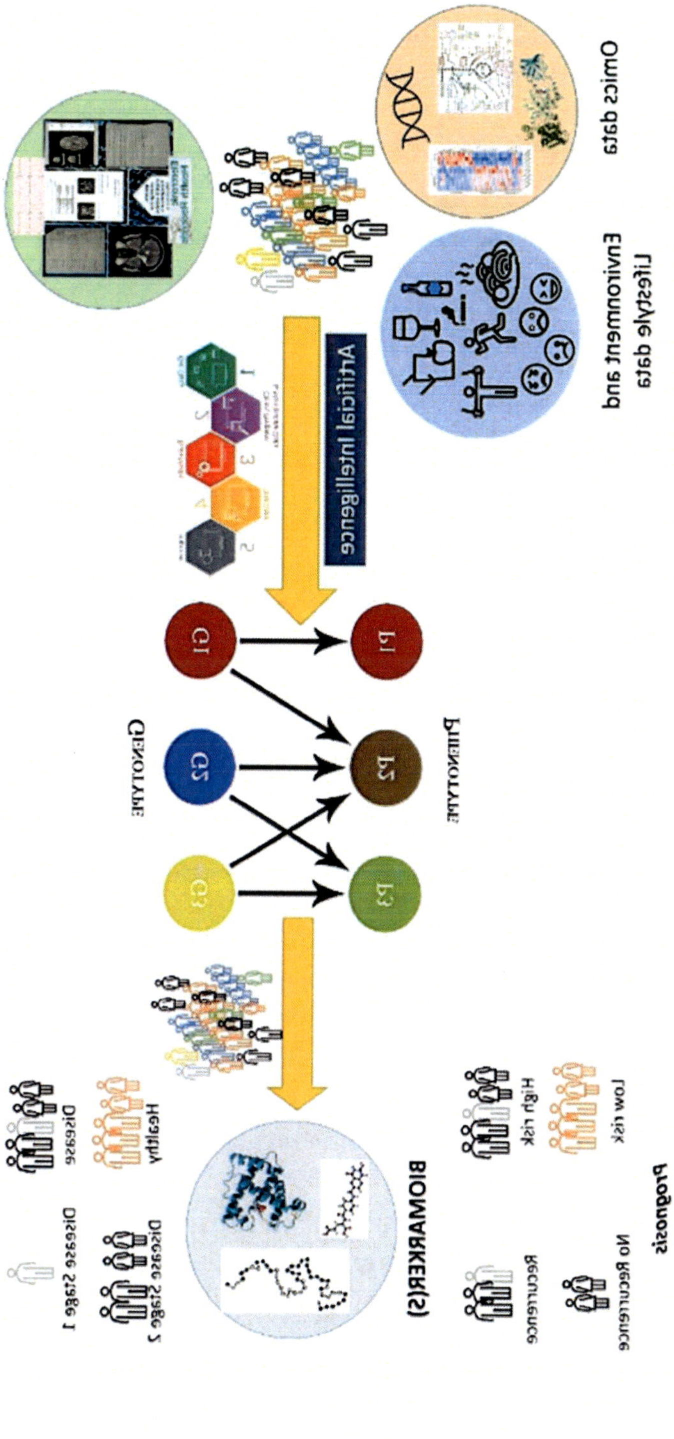

Fig. 4.1 Artificial intelligence can help in gaining insights from the heterogeneous datasets (clinical, omics, environmental, and lifestyle data), mapping genotype-phenotype relationships, and identifying novel biomarkers for patient diagnostics and prognosis against a specific disease

(predict occurrence, stage of disease), prognostic (patient susceptibility, disease recurrence, and overall patient survival), and other patient-based outcomes based on specific characteristics, succoring identification of clinically significant biomarkers (molecular markers).

4.2.1 Biomarker(s) for Diagnostics

Based on the diagnosis of a disease, clinicians may decide treatment pathway(s) for a patient with consideration to the patient's clinical history. In the past decade or so, efforts have been made to enable predictive diagnosis for diseases such as cancer, cardiac arrhythmia, gastroenterology, and other diseases. Data heterogeneity has been a major obstacle in the development of these early diagnostic applications. However, AI can aid in overcoming this challenge, as AI-trained algorithms can extract relevant knowledge from genomic and clinical datasets, such as disease-specific clinical molecular signatures or cohort-specific patterns. These genotype-phenotype relationships will render clinical management with an early diagnosis and patient stratification. In turn this should boost clinical decision-making among the available treatments, or mandatory treatment alterations, providing personalized bedside care to each patient.

The first set of AI-based applications in clinical diagnostics approved by the US FDA uses computer vision and is based on medical scans and/or pathological images: for example, the automated quantification of blood flow via cardiac MRI (Retson et al. 2019); determination of ejection fraction from ECG (Asch et al. 2019); mammography-based detection and quantification of breast densities (Le et al. 2019); detection of stroke, brain bleeds, and other conditions from CAT scans (FDA approves stroke-detecting AI software 2018); and diabetic retinopathy screening via dilated eye examination (van der Heijden et al. 2018). Furthermore, in cardiac arrhythmia, AI methods using deep neural networks can detect and classify arrhythmias, especially atrial fibrillation and cardiac contractile dysfunction (Tison et al. 2018; Attia et al. 2019; Hannun et al. 2019).

In addition to the conventional biomarkers, the focus is also on the exploration of digital biomarkers using hypothesis-driven approaches based on objective data, such as the data from wearable devices, adapting AI with IoT (Internet of Things) (Nam et al. 2019). Key applications are in development for digital biomarkers that might assist in early identification of spinal injuries and predict BP (blood pressure) status, which can facilitate early diagnosis and treatment of spinal and cardiovascular diseases, respectively (Guthrie et al. 2019; Nam et al. 2019).

4.2.2 Biomarker(s) for Disease Prognosis

Disease prognosis predominantly focuses on the prediction of susceptibility (risk assessment), recurrence, and survival of a patient. In terms of developing an AI application, these three terms can be defined in terms of probabilistic prediction.

Whereas risk assessment corresponds to developing a disease prior to its occurrence, recurrence is the possibility of regenerating the disease posttreatment, and survival is predicting an outcome post-diagnosis in terms of life expectancy, survivability, and/or disease progression. In the development of AI approaches for these prognostic predictors, we need to contemplate data elements besides clinical diagnosis. Therefore, amalgamating genomic factors, such as somatic mutation and/or expression of specific tumor proteins, with the clinical data shall strengthen the prognosis predictions.

In cancer, a prognosis usually involves varied subsets of biomarkers along with the clinical factors, the location and type of cancer, as well as the grade and size of the tumor (Edge and Compton 2010; Gress et al. 2017). For example, in ovarian cancer patients besides the physiological and genomic factors, CA125 (cancer antigen 125) protein estimation is used for risk assessment and recurrence prediction. Thus, considering the importance of personalized probabilistic predictions in cancer, the American Joint Committee on Cancer (AJCC) in 2016 illustrated the essential traits and guidelines that will help in developing prognostic predictive applications (Kattan et al. 2016).

Artificial neural networks (Rumelhart et al. 1986), decision trees (Quinlan 1986), genetic algorithms (Sastry et al. 2005), linear discriminant analysis (Duda et al. 2001), and nearest neighbor (Barber and Barber 2012) are the commonly used algorithms for developing prognostic predictive applications. Though in relation to identifying prognostic biomarkers via such applications, the predictive precision for a specific disease type is important for its adoption under clinical settings. For example, Oncotype DX is a prognostic test for breast cancer (ER+, HER2−) based on 21-gene panel scoring, which predicts recurrence and overall survival (McVeigh et al. 2014).

4.3 Role of Artificial Intelligence: Drug Discovery for Precision Medicine

Precision medicine is directed toward approaching a disease for treatment and prevention while including the genomic information, environmental factors, and lifestyle data of individuals. To achieve drug discovery in this scenario, drug discovery needs to be fast, efficient, and cost-effective. Drug discovery and development has always been a very sensitive and complex process, which time and again keeps challenging researchers as well as the pharmaceutical industry in terms of efficiency and R&D costs (Workman et al. 2019). To keep in pace with the approach of precision toward treatment and prevention of diseases, the drug discovery process requires an advancement with the help of the latest technologies. Drug development has largely benefited from incorporation of recent innovation technologies, and this has become utmost important in context to precision medicine. Precision medicine now marks a new relation between biomedical data and drug discovery as it provides us with an insight into mechanism and potential treatment options of a patient's disease. We will understand in this part how drug discovery process has been enabled by AI for effective and timely precision medicine delivery (Chen et al. 2018).

Precision medicine to its core is aimed at understanding the disease process in individual patients so that they can be divided into subgroups according to the different causes and influences of the disease. This promises delivery of more accurately personalized care to patients through drug discovery innovations and repurposing of drugs. Involvement of artificial intelligence is possible from the bench to the bedside as it can assist in the decision-making during various iterative phases of drug discovery, and it can help to determine the effective and appropriate therapy for a patient and, most importantly, assist in managing the clinical data generated and use it for future drug development (Duch et al. 2007; Vyas et al. 2018). In totality the drug discovery opportunities are completely different from the earlier times. Based on individual genetic variations and clinical, environmental, and lifestyle data, new therapeutic targets need to be located along with accelerated development of novel drugs or repurposed candidates and codevelopment of diagnostic tools for efficacious treatment of patient groups (Baronzio et al. 2015; Blasiak et al. 2020).

In the coming time, artificial intelligence is going to lead us toward fully addressing the human diseases through a thorough understanding of human biology. Incorporation of AI in healthcare will speed up the various processes involved in understanding the disease process in different patient subgroups and subsequent development of precision medicine. Various statistical and deep learning methods which rely on data interpretation will pave a way for diagnosis and classification of diseases and disease subtype among patients. The use of machine learning, clustering, and feature finding methods could be helpful in the discovery of disease targets in an accurate and fast manner (Sellwood et al. 2018; Mak and Pichika 2019). The use of neural networks, big data, and data mining algorithms along with enhanced statistical analysis on experimental data will enhance our ability for de novo drug design. Based on various genetic makers and improved patient information, repurposing and combination therapies of drugs will improve the area of precision medicine.

4.3.1 Drug Discovery Process

In order to approach precision medicine delivery by utilizing artificial intelligence, the drug discovery process itself needs to be enabled by artificial intelligence techniques at various stages. Drug discovery is an iterative process, which requires continuous inputs and feedbacks at each step for better drug development. It can significantly benefit from utilization of various AI-based techniques during various stages of drug discovery. These techniques have an important role to play to enable the timely incorporation of accurate inputs at every step of drug discovery especially in the case of precision medicine where we have a variety of data for the same disease pertaining to different subgroups. Drug discovery process begins with identification and validation of a target molecule, followed by identification of a compound with a promising biological activity. Identification of a potential compound itself is an iterative and multistage process (Grys et al. 2017; Jiang et al. 2017; Labovitz et al. 2017; Zhu 2020). It begins with identification of a "hit" using various computational screening techniques, followed by "lead" identification, which is achieved by screening of hits in various cell-based assays and animal models to access the safety and efficacy of the lead molecule. Hit to lead identification process is a highly iterative process during which hits are continuously modified to generate lead molecules with an improved activity and selectivity toward target molecules and reduced toxicity. During the process of lead generation, there is a scope of exploring the chemical space surrounding the hit molecules by developing analogues. This process is called hit expansion, and medicinal chemists often exploit binding site information for the development of better promising analogues, where the binding site information. The most promising compounds identified computationally need to be synthesized for further experimental evaluation (in vitro and in vivo analysis) (Yuan et al. 2011; Zhu 2013; Fleming 2018). In fact the lead identification and optimization step is the most time-consuming and crucial step in drug discovery. Experimental evaluation is followed by preclinical and clinical trials. Let us now understand the role of artificial intelligence as applicable in different stages of drug discovery as depicted in Fig. 4.2 (Anderson 2012; Hall et al. 2012).

4.3.2 Understanding the Disease Process and Target Identification

A very strong determinant of success of a drug discovery process is, firstly, the detailed understanding of the disease process and, secondly, drug-target identification and validation. Artificial intelligence enables the evaluation of vast amount of structural and functional genomic data, proteomic data, and in vitro and in vivo assays. Artificial intelligence algorithms also analyze large amount of research data available at various private and public platforms to help up better understand the disease process and pathways associated, which was not possible earlier. Some AI-based platforms have already been developed, which utilize extensive literature information, genomic data, disease-associated data, and other relevant data for target identification and validation in days rather than months, e.g., Open Targets, IBM Watson for drug discovery, Benevolent Platform, etc.

4.3.3 Identification of Hit and Lead

The process of compound screening and lead optimization is the most time-consuming and costly step in the entire drug discovery process. The process involves selection of candidate using combinatorial chemistry, high-throughput screening, and virtual screening. The implementation of artificial intelligence to explore the chemical space makes it possible to identify novel and high-quality molecules with a reduced cost and time. The idea is to search for bioactive compounds by using

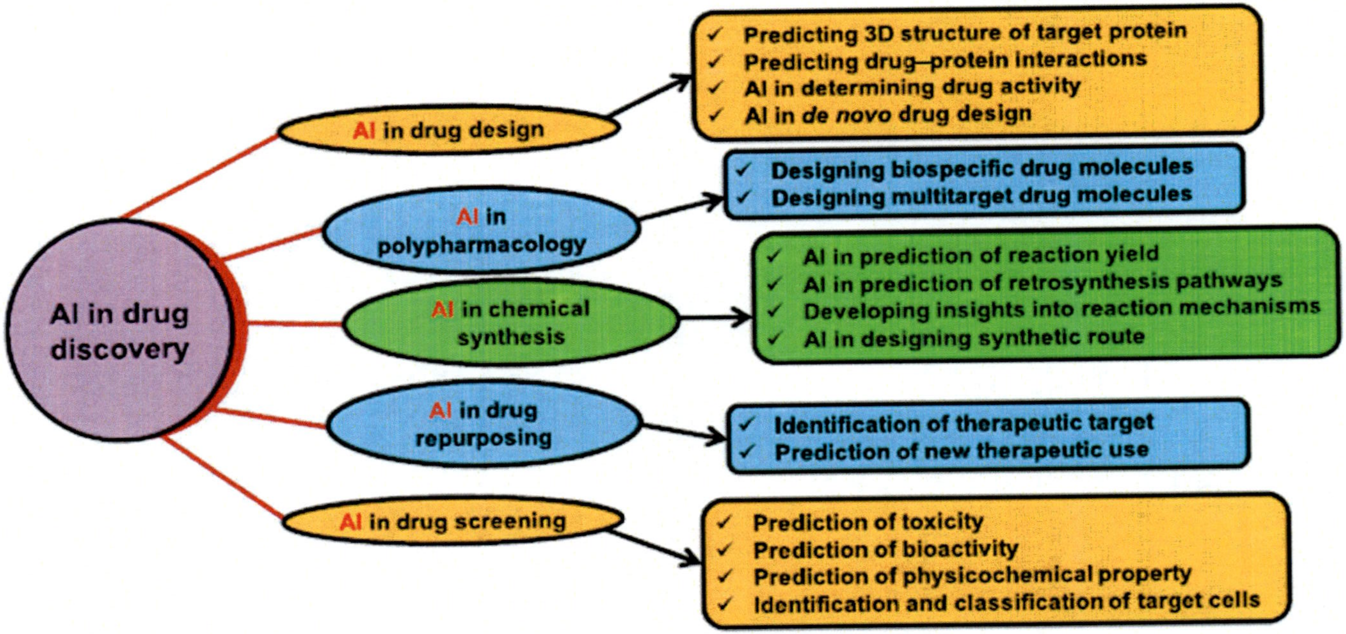

Fig. 4.2 Application of artificial intelligence in various steps of drug discovery process (Paul et al. 2020)

AI-based virtual screening to help select appropriate molecules for further testing. This can be done by using publicly available chemical spaces including PubChem, ChemBank, DrugBank, and ChemDB. Some molecules can also be extracted from mining the research literature using AI-based techniques, which can be further modified to develop some workable analogues. To speed up the initial phase of drug screening, potential lead molecules can be efficiently screened by medicinal chemists by application of artificial neural networks, support vector machines, Bayesian classifiers, and k-nearest neighbors and other algorithms on millions of compounds.

AI-based systems can help to reduce the number of compounds for synthesis and subsequent testing in vitro and in vivo by screening only the most promising compounds and hence help in reduction of R&D expenditure by decreasing the dropout rate. The compounds can be filtered during the screening process based on predicted pharmacokinetic properties, bioactivity, and toxicity. Several programs have been used to predict the lipophilicity, solubility, and drug-target interactions. Some of the examples are ALGOPS; neural networks based on the ADMET predictor, which predicts the lipophilicity and solubility; graph-based convolutional neural networks (CVNN), which predicts solubility of molecules; and ChemMapper and the similarity ensemble approach (SEA), for predicting drug-target interactions to access the advanced based on input features. Several AI-based approaches predict the toxicity of the compound based on similarities among compounds. Some major biopharma companies working in different areas such as cardiovascular diseases and fibrosis have started collaborating with AI-based companies for de novo design of molecules, antibodies, DNA, and peptides. One of the successful cases is a de novo designed compound using AI, which was developed in just 25 days by Insilico Medicine and was found to be 15 times faster than traditional biopharma process (Mayr et al. 2016; Segler et al. 2018).

It has already been established that AI techniques can help to speed up and increase the success rates in drug development, but it is always recommended to validate the AI techniques before applying to the drug development process.

4.3.4 Synthesis of Compounds

The synthesis of chosen molecules is the most important step in the drug development process. AI is valuable at this stage too, owing to its ability to deduce the optimal synthetic route and to prioritize molecules based on the ease of synthesis (Alanine et al. 2012; Okafo et al. 2018). The synthesis of compounds begins with fragmenting a target compound into building blocks and then establishing an optimal reaction process for synthesis of the compound. The optimization of reaction is the most challenging step with chances of failure of the rate of synthesis. AI would aid in predicting the best sought-after reactions by predicting and working upon the cause of high failure in this process. Artificial intelligence can be used to automate chemical synthesis with minimal manual operation using synthesis robots combined with artificial intelligence. Currently, for the selection of the synthesis route, various systems are available to assist the chemists such as CAOCS (computer-aided organic compound synthesis) (Paul et al. 2020). From a group of building blocks, filtering out only the most promising ones for synthesis of target compounds using well-known reactions can be achieved by using an AI platform named 3 N-MCTS. Computer-aided organic compound synthesis using 3 N-MCTS is achieved by using three different deep neural networks with Monte Carlo tree search.

4.3.5 Predicting the Drug-Target Interactions Using AI

Assignment of a correct target to a drug molecule is essential for a successful treatment. It is very vital to predict the target protein structure for selective targeting of the disease. AI can assist in exploring the structural and chemical environment of the target and designing the molecules exhibiting physically and chemically complementarity with the binding site (Paul et al. 2020). This will help to select only highly effective compounds with safety considerations for further synthesis and production. Drug-target interactions have been very well explained by lock-and-key model, where the target is the lock and the drug molecule is the key. AI with the help of its highly predictive algorithms and data analysis techniques can also be useful to find out new locks (drug targets) for the already existing keys (drugs). Some tools based on AI have already been developed to assist in the process, e.g., AlphaFold, NN-based methods, etc. The success of a therapy is highly dependent on drug-protein interactions. The understanding and accurate prediction of drug-target interactions play an important role to improve the efficacy of the drug and explore more molecules for drug repurposing. Various AI-based methods have already been developed such as SVM-based model, which was used to predict the drug-target interactions after being trained on 15,000 interactions. AI-based prediction algorithms are also capable of assisting in repurposing of existing drugs and avoiding polypharmacology (Paul et al. 2020). Drug repurposing is a very efficient and cost-effective method as the repurposed drug qualifies directly for Phase II clinical trials. Thus, R&D expenditure is reduced because, in comparison with the launch of a new drug, relaunching an existing drug costs very less.

4.3.6 Artificial Intelligence in Clinical Trials

Clinical trial is a very important stage of drug discovery which can be 6–7 years long and requires a substantial financial investment. Despite such a big investment in terms of time and money, only one out of ten molecules on an average becomes successful, which is a massive loss to the industry. During the conduct of the trials, multiple factors such as from inappropriate patient selection, shortage of technical requirements, and poor infrastructure contribute to the failure (Bain et al. 2017).

Patient selection in various phases of clinical trials is a very crucial process. During the clinical trial process, the therapeutic responses of patients are very uncertain. For a predictable, accurate, and quantifiable assessment of the response data, investigation of the relationship between human-relevant biomarkers and in vitro phenotypes is essential. The recruitment of patients for Phases II and III clinical trial stages can be assisted by AI approaches for identification and prediction of human-relevant biomarkers of disease (Perez-Gracia et al. 2017). An efficient and ideal AI tool would be one which can identify the gene target and predict the effect of the molecule designed in addition to recognizing the disease in the patient. The use of AI-based predictive modeling would increase the success rate in clinical trials. It has been observed that the failure of 30% of the clinical trials is due to dropout of patients from clinical trials. This leads to wastage of time, money, and incomplete data and creates a need of additional recruiting requirements for the completion of the trial, thus increasing the cost further. Association and adherence of the patients can be increased by close monitoring of the patients and developing methods which can help them to easily follow the desired protocol of the clinical trial.

4.3.7 Drug Repurposing

The repurposing of drugs has become more promising with the inclusion of AI in drug discovery. Application of an existing therapeutic to a new disease is a cost-effective and fast drug discovery application because the new drug is qualified to go directly to Phase II trials for a different indication without having to pass through Phase I clinical trials and toxicology testing again (Corsello et al. 2017). AI-based deep neural networks and reinforcement learning are used to identify drug molecules exhibiting certain patterns in the molecular structure that can suggest their use in other new diseases. AI-based methods can efficiently mine the research data from literature and help to identify certain compounds, which can be repurposed for other diseases. In the later stages AI-based tools are efficient in market prediction and analysis. In some cases AI-based nanorobots are also being used for efficient delivery of drugs (Hernandez et al. 2017).

4.3.8 Some Examples of AI and Pharma Partnerships

With the promise of providing better healthcare to the patients and as a way forward toward precision medicine, a new sync has started developing between pharmaceutical companies and AI companies. Pharmaceutical companies, hospitals, and other healthcare agencies have started to work along with AI companies in the hopes of developing better healthcare tools. The joint ventures with the aim of improving the diagnosis through biomarkers, target identification, and novel drug design have already begun through tool development, data analysis, and data exchange with the aim of (Mak and Pichika 2019). Various partnerships between pharmaceutical industries and AI companies on a global scale were recently developed as depicted in Fig. 4.3.

AI has shown great promise in rapidly evolving drug design process through accurate and fast predictions of the existing as well as newly designed compounds

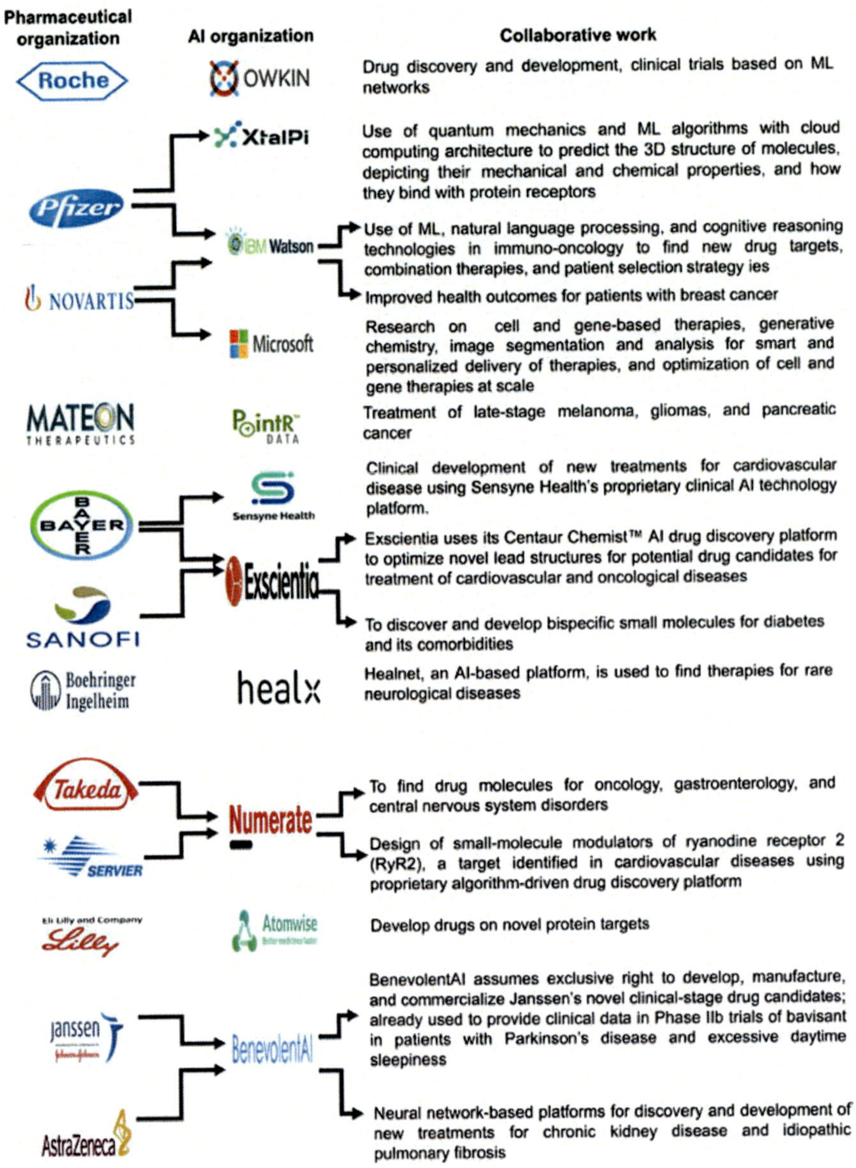

Fig. 4.3 Some examples of pharmaceutical companies collaborating with artificial intelligence (AI) organization for healthcare improvements in the field of oncology, cardiovascular diseases, and central nervous system disorders (Paul et al. 2020)

and better exploration of drug targets. These advancements will serve as the most important contributing factors for the betterment of healthcare services, improvement in terms of efficiency in clinical trials, enhancement in stratified medicine, and more. Presently, drug discovery costs billions and takes around 12–15 years to complete. This trend is unsustainable in this fast-changing world, and positive change is extremely essential (Paul et al. 2020). These collaborations will help not only to improve upon the existing design space but also to discover and explore rare molecules that have properties of extreme importance, which were impossible to identify by solely relying on conventional methods. In the present scenario, it is a challenge to develop a drug especially while including the genetic/genomic information, environmental factors, and lifestyle of individuals for precision medicine development. It takes thousands of studies to analyze known side effects and unknown interactions. However, once available, such an AI algorithm approach would prove invaluable in further hastening drug development efforts. AI will revolutionize how drugs are discovered and will reinvent the pharmaceutical industry along with precision medicine.

4.4 Precision Medicine and Artificial Intelligence: Hopes and Challenges

Artificial intelligence-based approaches are in forefront for biomarker and drug discovery, leading to therapeutic interventions. It is already improving the clinical care scenario, and quite a few successful examples are there for complex diseases such cancer and cardiovascular disease. However, there are quite a few technical challenges such as reliability and reproducibility that need to be addressed. This often arise with difference in the protocols being used to collect data and/or algorithms for analysis. Computationally, these are eliciting research issues, as optimization is foreseen between the performance and interpretability of AI engendered learning. This includes implementing algorithms centered on the requirements of the clinical care providers toward a particular disease and patients, offering the motivation to substantiate the outcomes.

References

Alanine A et al (2012) Lead generation—enhancing the success of drug discovery by investing in the hit to Lead process. In: Combinatorial chemistry & high throughput screening. Bentham Science, Sharjah. https://doi.org/10.2174/1386207033329823

Anderson AC (2012) Structure-based functional design of drugs: from target to lead compound. Methods Mol Biol 823:359–366. https://doi.org/10.1007/978-1-60327-216-2_23

Asch FM et al (2019) Accuracy and reproducibility of a novel artificial intelligence deep learning-based algorithm for automated calculation of ejection fraction in echocardiography. J Am Coll Cardiol 73:1447. https://doi.org/10.1016/s0735-1097(19)32053-4

Atkinson AJ et al (2001) Biomarkers and surrogate endpoints: preferred definitions and conceptual framework. Clin Pharmacol Ther 69:89–95. https://doi.org/10.1067/mcp.2001.113989

Attia ZI et al (2019) Screening for cardiac contractile dysfunction using an artificial intelligence–enabled electrocardiogram. Nat Med 25:70–74. https://doi.org/10.1038/s41591-018-0240-2

Bain EE et al (2017) Use of a novel artificial intelligence platform on mobile devices to assess dosing compliance in a phase 2 clinical trial in subjects with schizophrenia. JMIR Mhealth Uhealth 5:e18. https://doi.org/10.2196/mhealth.7030

Barber D, Barber D (2012) Nearest neighbour classification. In: Bayesian reasoning and machine learning. Cambridge University Press, London. https://doi.org/10.1017/cbo9780511804779.019

Baronzio G, Parmar G, Baronzio M (2015) Overview of methods for overcoming hindrance to drug delivery to tumors, with special attention to tumor interstitial fluid. Front Oncol 5:115. https://doi.org/10.3389/fonc.2015.00165

Biomarker Working Group FDA NIH (2016) BEST (biomarkers, EndpointS, and other tools). FDA-NIH Biomarker Working Group, Silver Spring

Blasiak A, Khong J, Kee T (2020) CURATE.AI: optimizing personalized medicine with artificial intelligence. In: SLAS technology. SAGE, Thousand Oaks, pp 95–105. https://doi.org/10.1177/2472630319890316

Burki TK (2017) Defining precision medicine. Lancet Oncol 18(12):e719. https://doi.org/10.1016/S1470-2045(17)30865-3

Chen H et al (2018) The rise of deep learning in drug discovery. Drug Discov Today 23:1241–1250. https://doi.org/10.1016/j.drudis.2018.01.039

Corsello SM et al (2017) The drug repurposing hub: a next-generation drug library and information resource. Nat Med 23:405–408. https://doi.org/10.1038/nm.4306

Deliberato RO, Celi LA, Stone DJ (2017) Clinical note creation, binning, and artificial intelligence. JMIR Med Inform 5:e24. https://doi.org/10.2196/medinform.7627

Della-Morte D, Pacifici F, Rundek T (2016) Genetic susceptibility to cerebrovascular disease. Curr Opin Lipidol 27:187–195. https://doi.org/10.1097/MOL.0000000000000275

Deng X, Nakamura Y (2017) Cancer precision medicine: from cancer screening to drug selection and personalized immunotherapy. Trends Pharmacol Sci 38:15–24. https://doi.org/10.1016/j.tips.2016.10.013

Duch W, Swaminathan K, Meller J (2007) Artificial intelligence approaches for rational drug design and discovery. Curr Pharm Des 13:14. https://doi.org/10.2174/138161207780765954

Duda RO, Hart PE, Stork DG (2001) Pattern classification. Wiley, New York

Edge SB, Compton CC (2010) The american joint committee on cancer: the 7th edition of the AJCC cancer staging manual and the future of TNM. Ann Surg Oncol 17:1471–1474. https://doi.org/10.1245/s10434-010-0985-4

FDA approves stroke-detecting AI software (2018) FDA approves stroke-detecting AI software. Nat Biotechnol 36:290

FDA-NIH Biomarker Working Group (2016) BEST (biomarkers, EndpointS, and other tools) resource [internet], updated, Sept 25

Fleming N (2018) How artificial intelligence is changing drug discovery. Nature 557:55–57. https://doi.org/10.1038/d41586-018-05267-x

Gress DM et al (2017) Principles of cancer staging. In: AJCC cancer staging manual. Springer, Cham. https://doi.org/10.1007/978-3-319-40618-3_1

Grys BT et al (2017) Machine learning and computer vision approaches for phenotypic profiling. J Cell Biol 216:65–71. https://doi.org/10.1083/jcb.201610026

Guide Y, Conditions UG (2015) What is the difference between precision medicine and personalized medicine? What about pharmacogenomics? Genetics Home Reference

Guthrie NL et al (2019) Emergence of digital biomarkers to predict and modify treatment efficacy: machine learning study. BMJ Open 9:e030710. https://doi.org/10.1136/bmjopen-2019-030710

Hall DR et al (2012) Hot spot analysis for driving the development of hits into leads in fragment-based drug discovery. J Chem Inf Model 52(1):199–209. https://doi.org/10.1021/ci200468p

Hannun AY et al (2019) Cardiologist-level arrhythmia detection and classification in ambulatory electrocardiograms using a deep neural network. Nat Med 25:65–69. https://doi.org/10.1038/s41591-018-0268-3

Hauser A et al (2017) National molecular surveillance of recently acquired HIV infections in Germany, 2013 to 2014. Eurosurveillance 22:30436. https://doi.org/10.2807/1560-7917.ES.2017.22.2.30436

Hernandez JJ et al (2017) Giving drugs a second chance: Overcoming regulatory and financial hurdles in repurposing approved drugs as cancer therapeutics. Front Oncol 7:273. https://doi.org/10.3389/fonc.2017.00273

Jiang F et al (2017) Artificial intelligence in healthcare: past, present and future. *Stroke Vasc Neurol* 2:230–243. https://doi.org/10.1136/svn-2017-000101

Joyner MJ, Paneth N (2019) Promises, promises, and precision medicine. J Clin Investig 129:946–948. https://doi.org/10.1172/JCI126119

Kattan MW et al (2016) American joint committee on cancer acceptance criteria for inclusion of risk models for individualized prognosis in the practice of precision medicine. CA Cancer J Clin 66 (5):370–374. https://doi.org/10.3322/caac.21339

König IR et al (2017) What is precision medicine? Eur Respir J 50:1700391. https://doi.org/10.1183/13993003.00391-2017

Labovitz DL et al (2017) Using artificial intelligence to reduce the risk of nonadherence in patients on anticoagulation therapy. Stroke 48:1416–1419. https://doi.org/10.1161/STROKEAHA.116.016281

Le EPV et al (2019) Artificial intelligence in breast imaging. Clin Radiol 74:357–366. https://doi.org/10.1016/j.crad.2019.02.006

Mak KK, Pichika MR (2019) Artificial intelligence in drug development: present status and future prospects. Drug Discov Today 24(3):773–780. https://doi.org/10.1016/j.drudis.2018.11.014

Mayr A et al (2016) DeepTox: toxicity prediction using deep learning. Front Environ Sci. https://doi.org/10.3389/fenvs.2015.00080

McVeigh TP et al (2014) The impact of Oncotype DX testing on breast cancer management and chemotherapy prescribing patterns in a tertiary referral centre. Eur J Cancer 50:2763. https://doi.org/10.1016/j.ejca.2014.08.002

Nam KH et al (2019) Internet of things, digital biomarker, and artificial intelligence in spine: current and future perspectives. Neurospine 16:705–711. https://doi.org/10.14245/ns.1938388.194

Okafo G et al (2018) Adapting drug discovery to artificial intelligence. Drug Target Rev

Pacanowski M, Huang SM (2016) Precision medicine. Clin Pharmacol Ther 99:124–129. https://doi.org/10.1002/cpt.296.

Paul D et al (2020) Artificial intelligence in drug discovery and development. Drug Discov Today 26:80–93. https://doi.org/10.1016/j.drudis.2020.10.010

Perez-Gracia JL et al (2017) Strategies to design clinical studies to identify predictive biomarkers in cancer research. Cancer Treat Rev 53:79. https://doi.org/10.1016/j.ctrv.2016.12.005

Pinto AC et al (2013) Trastuzumab for patients with HER2 positive breast cancer: delivery, duration and combination therapies. Breast 22:152. https://doi.org/10.1016/j.breast.2013.07.029

Quinlan JR (1986) Induction of decision trees. Mach Learn 1:81–106

Retson TA et al (2019) Machine learning and deep neural networks in thoracic and cardiovascular imaging. J Thorac Imaging 34:192. https://doi.org/10.1097/RTI.0000000000000385

Rumelhart DE, Hinton GE, Williams RJ (1986) Learning representations by back-propagating errors. Nature 323(6088):533–536. https://doi.org/10.1038/323533a0

Sankar PL, Parker LS (2017) The precision medicine initiative's all of us research program: an agenda for research on its ethical, legal, and social issues. Genet Med 19:743. https://doi.org/10.1038/gim.2016.183

Sastry K, Goldberg D, Kendall G (2005) Genetic algorithms. In: Search methodologies: introductory tutorials in optimization and decision support techniques. Springer, New York, pp 97–125. https://doi.org/10.1007/0-387-28356-0_4

Scheen AJ (2016) Precision medicine: the future in diabetes care? Diabetes Res Clin Pract 117:12–21. https://doi.org/10.1016/j.diabres.2016.04.033

Segler MHS, Preuss M, Waller MP (2018) Planning chemical syntheses with deep neural networks and symbolic AI. Nature 555:604. https://doi.org/10.1038/nature25978

Sellwood MA et al (2018) Artificial intelligence in drug discovery. Future Med Chem 10:2025. https://doi.org/10.4155/fmc-2018-0212

Slikker W (2018) Biomarkers and their impact on precision medicine. Exp Biol Med 243 (3):211–212. https://doi.org/10.1177/1535370217733426

Tison GH et al (2018) Passive detection of atrial fibrillation using a commercially available smartwatch. JAMA Cardiol 3:409. https://doi.org/10.1001/jamacardio.2018.0136

van der Heijden AA et al (2018) Validation of automated screening for referable diabetic retinopathy with the IDx-DR device in the Hoorn diabetes care system. Acta Ophthalmol 96(1):63–68. https://doi.org/10.1111/aos.13613

Vyas M et al (2018) Artificial intelligence: the beginning of a new era in pharmacy profession. Asian J Pharm 12:72–76

Wang RF, Wang HY (2017) Immune targets and neoantigens for cancer immunotherapy and precision medicine. Cell Res 27:11–37. https://doi.org/10.1038/cr.2016.155

Weil AR (2018) Precision medicine. Health Aff 37:687–687. https://doi.org/10.1377/hlthaff.2018.0520

Workman P, Antolin AA, Al-Lazikani B (2019) Transforming cancer drug discovery with big data and AI. Expert Opin Drug Discov 14:11. https://doi.org/10.1080/17460441.2019.1637414

Yuan Y, Pei J, Lai L (2011) LigBuilder 2: A practical de novo drug design approach. J Chem Inf Model 51:1083–1091. https://doi.org/10.1021/ci100350u

Zhu T et al (2013) Hit identification and optimization in virtual screening: practical recommendations based on a critical literature analysis. J Med Chem 56(17):6560–6572. https://doi.org/10.1021/jm301916b

Zhu H (2020) Big data and artificial intelligence modeling for drug discovery. Annu Rev Pharmacol Toxicol 60:573–589. https://doi.org/10.1146/annurev-pharmtox-010919-023324

Transfer Learning in Biological and Health Care

5

Abstract

Transfer learning is the advancement over conventional machine learning in which we transfer the newly obtained knowledge to existing knowledge. Traditional machine learning makes a basic assumption that the distribution of training data and testing data should be the same. But in numerous real-world cases, this identical-distribution assumption of training data and testing data does not hold at all. For example, suppose if we have a model to recognize a face from an image in traditional machine learning, we cannot retrain this model to detect tumors in the brain because they belong to a different domain. But using transfer learning, we can retrain this model to detect tumors as well. The identical-distribution assumption might be violated in cases where data from one new domain comes, while there are only available labeled data from a similar other domain. Labeling the new data in the old domain can be costly for any organization, and it is also inappropriate to throw away the newly obtained data just because it is from another domain.

In this chapter, we retrained various state-of-the-arts convolutional deep learning models using transfer learning by supplying the data related to brain tumors using transfer learning technique.

Keywords

Machine learning · Keras · Tumors

5.1 Introduction

After computers came into existence in the 1950s and 1960s, various algorithms were constructed and developed, which enable us to model and analyze a large amount of data. This leads to the existence of initial machine learning techniques.

A. Saxena, S. Chandra, *Artificial Intelligence and Machine Learning in Healthcare*,
https://doi.org/10.1007/978-981-16-0811-7_5

Three branches of machine learning emerged in the very beginning. These are classical works that include neural networks by Rosenblatt (Rosenblatt 1962), statistical methods by Nilsson (Nilsson and Machines 1965), and symbolic learning by Hunt et al. (Hunt et al. 1966).

Over the years, these three techniques were enhanced to construct improved techniques (Michie et al. 1994): pattern recognition or statistical methods, such as the Bayesian classifier and k-nearest neighbors classifier; inductive learning, such as decision trees; and artificial neural networks (ANN), such as the multilayered feedforward neural network (MLP) including back propagation (Kononenko 2001).

Machine learning algorithms discover patterns in data that are finding a predictive relationship between different variables. Mostly, we can say that finding where the probability of mass concentrates on the joint distribution of all the observations observed (LeCun et al. 2015).

Earlier machine learning algorithms have limited abilities to process data and signal in their natural/raw form. The development of a machine learning system or pattern-recognition system requires domain expertise and strict engineering to design a feature extraction algorithm from scratch that transformed the natural/raw data into a feature vector or suitable internal representation of the data from which a machine learning classifier could detect or classify patterns based on the input provided (LeCun et al. 2015).

To overcome the earlier machine learning limitations, artificial intelligence community introduces deep learning.

Deep learning has turned out to be outperforming earlier machine learning techniques in terms of finding intricate patterns within high-dimensional data, and, therefore, it has applications in many different domains such as fields of science, business, and government (LeCun et al. 2015).

Using deep learning one can solve various complex problems with ease as compared to traditional machine learning approaches; such complex problems include computer vision, natural language processing, signal processing, anomaly detection system, recommendation system, and so on.

Deep learning outperforms earlier machine techniques in the field of image recognition and speech recognition. It has been found that deep learning also outperforms machine learning in the field of bioinformatics such as predicting the activity of possible drug molecules, analyzing a large amount of particle accelerator data to provide its analysis, reconstructing brain circuitry, and predicting the effects of various mutations in noncoding DNA region of the gene of disease samples (LeCun et al. 2015).

More surprisingly, various tasks related to natural language understanding are sentiment analysis, topic classification, language translation, and question answering that show the most promising results when done using deep learning algorithms (LeCun et al. 2015).

Several machine learning techniques work well only under common assumptions, i.e., the training and testing samples should be obtained from the same distribution and also with the same feature space. When this distribution changes, then most of the statistical models require rebuilding again from scratch by using newly obtained training data. But in numerous real-world applications, it is expensive or impossible to recollect all the training data and then rebuild the models from scratch because of many reasons such as resource limitation, lack of computational power, and others (Pan and Yang 2010).

In these cases, we use a technique called as transfer learning or knowledge transfer to retrain our existing trained model on newly acquired data instead of building the entire model again (Pan and Yang 2010).

Using transfer learning has its own advantages such as follows:

- By using this, there might be a boost in model baseline performance.
- Since we do not need to create an entire model from scratch, our model development time is reduced significantly.
- Training with a small number of training samples can give you better results as compared to the model made from scratch.

In the field of deep learning, companies such as Google, Facebook, and Microsoft and some researchers regularly contributed to giving us high-performance and optimized convolutional deep learning models that are trained on millions of image samples belonging to over 1000 different classes containing over a billion trainable parameters. Training of such models requires a huge amount of computational power, which is quite impossible to arrange for individuals or independent researchers.

Transfer learning in the perspective of deep learning is defined as fine-tuning of weights and biases of an existing trained model by retraining it using the newly collected data. Training a model using the transfer learning technique inherits the characteristic of the features of the previous data on which it was trained on a model as well as the features of the newly obtained data.

5.2 Methodology

Retraining an existing pretrained model using transfer learning is almost similar to developing the model from scratch. The only difference in this is that we do not need to design the model from scratch. We can reuse the existing model and their respective weights.

Steps to retrain a pretrained model include data curation, data loading and preprocessing, loading the existing model and its weights, training, and testing.

We are using Keras on the TensorFlow back end for loading images and pretrained model, training, and testing. Pretrained model and their respective weights were loaded using Keras application API. Training, validation, and testing images were loaded, preprocessed, and augmented using Keras Image Generator API.

We have use Google Colaboratory platform to do all the abovementioned tasks.

5.2.1 Dataset Curation

We have curated our dataset from an open-source repository known as Kaggle; see Table 5.1. We divided our complete data into three individual parts, namely, training dataset on which we perform training, validation dataset on which we perform cross-validation, and testing dataset on which we perform testing; see Table 5.2.

Our training dataset consists of a total of 193 brain CT scan images, out of which 119 image samples are of patients having a tumor in their brain and the rest 74 image samples are of healthy patients.

Our validation dataset consists of a total of 50 brain CT scan images, out of which 31 image samples are of patients having a tumor in their brain and 19 image samples are of healthy patients.

Our testing dataset consists of a total of ten brain CT scan images, out of which five image samples are of patients having a tumor in their brain and five image samples are of healthy patients.

5.2.2 Data Loading and Preprocessing

We are using images as a training sample, in terms of computer interpretation of the image in a 2D matrix, and each cell contains a value ranging 0–255 based on its intensity called pixel.

The performance of deep learning algorithms is significantly reduced in the presence of outliers. When we supply the data containing a significant amount of outliers, it causes the problem. So, we need to scale or normalize our data to get rid of outliers.

We scale our pixel values such that every pixel data is ranging from 0 to 1 but dividing the current pixel intensity with 255, which is a maximum value of single pixel.

Table 5.1 *Description of dataset*: we have in total 253 brain MRI images out of which 155 are having tumor and 98 are normal

Category	Quantity	Total
Tumor-containing MRI scans	155	253
Healthy MRI scans	98	

Table 5.2 *Description of dataset type*: we have in total 253 brain MRI images. We split our whole dataset into three different parts: training, validation, and testing dataset

Dataset type	Number of images of tumor patients	Number of images of normal patients	Total
Training	119	74	253
Validation	31	19	
Testing	5	5	

We also use image augmentation which increases our model generality and robustness to unknown sample images. We are using shearing, rotation, flipping, and skew image augmentation.

We are using Keras Image Generator API to load, scale, and image augmentations of our database.

5.2.3 Loading Transfer Learning Models

5.2.3.1 VGG-16

VGG-16 is developed by Andrew Zisserman and Simonyan in 2014 at the Visual Geometry Group Lab of Oxford University (Karen Simonyan and Andrew Zisserman 2018). This model achieved top five test accuracy of 92.7% on the ImageNet dataset. This ImageNet dataset contains 14 million images belonging to 1000 different classes.

The input image size of the VGG-16 model is 224 × 224 × 3, i.e., 224 image height × 224 pixel image width × 3 channels (RGB).

We use Keras application VGG-16 class to create our transfer learning model using "ImageNet" weights along with some modifications as described in Fig. 5.1. We retrained the VGG-16 model with Adam optimizer along with exponential learning rate decay and binary cross-entropy loss function.

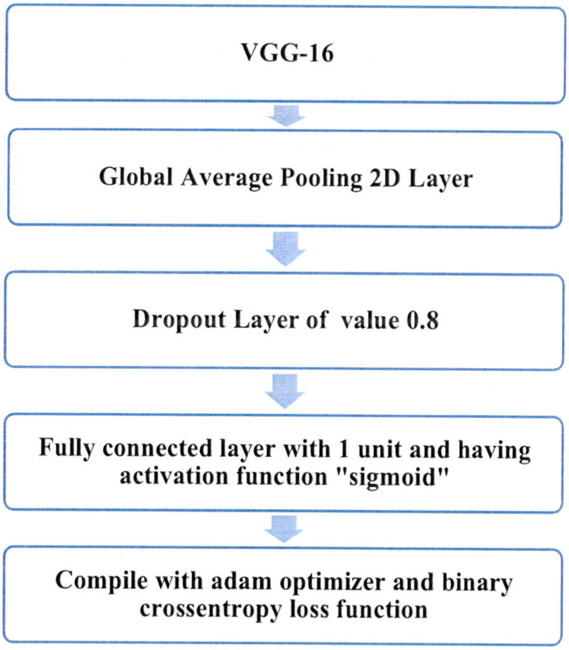

Fig. 5.1 Modified VGG-16 model

5.2.3.2 EfficientNet

EfficientNet is developed by Mingxing Tan and Quoc V. Le in ICML 2019 (Tan and Le 2019). Efficient Net achieved top-1 accuracy of 84.4 and top-5 accuracy of 97.1% on the ImageNet dataset. The complete model consists of 66 million parameters in total.

The input image size of the EfficientNetB4 model is 229 × 229 × 3, i.e., 229 image height × 229 pixel image width × 3 channels (RGB).

We use Keras application EfficientNetB4 class to create our transfer learning model using "ImageNet" weights along with some modifications as described in Fig. 5.2. We retrained the EfficientNetB4 model with Adam optimizer along with exponential learning rate decay and binary cross-entropy loss function.

5.2.3.3 Inception-ResNet-V2

Inception-ResNet-V2 is developed by the team of Christian Szegedy, Sergey Ioffe, Alex Alemi, and Vincent Vanhoucke in 2016 (Szegedy et al. 2016).

Inception-ResNet-V2 is a convolutional neural network (CNN) that achieved a top-1 accuracy of 80.4% and top-5 accuracy of 95.3% in the ILSVRC image classification. Inception-ResNet-V2 is a version of Inception-V3 network, which implements some ideas from Microsoft's ResNet network.

The input image size of the Inception-ResNet-V2 model is 229 × 229 × 3, i.e., 229 image height × 229 pixel image width × 3 channels (RGB).

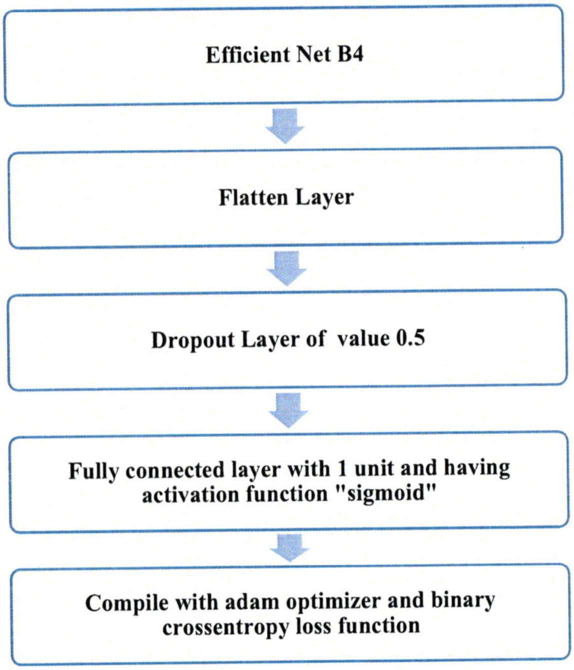

Fig. 5.2 Modified EfficientNetB4 model

Fig. 5.3 Modified Inception-ResNet-V2 model

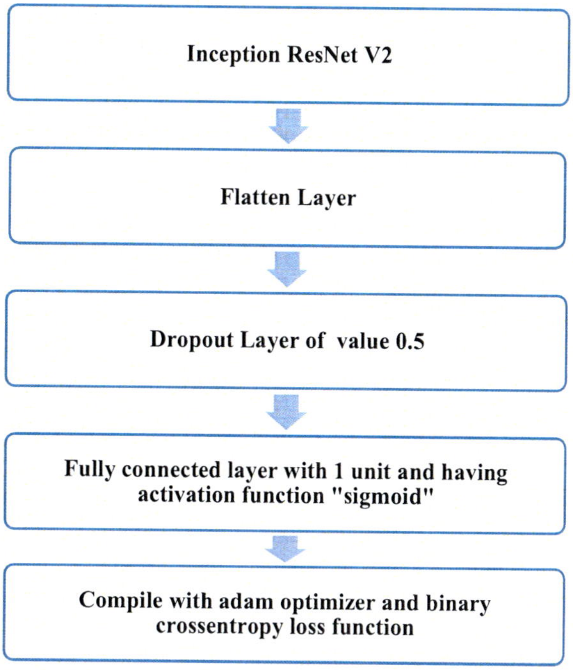

We use Keras application Inception-ResNet-V2 class to create our transfer learning model using "ImageNet" weights along with some modifications as described in Fig. 5.3. We retrained the Inception-ResNet-V2 model with Adam optimizer along with exponential learning rate decay and binary cross-entropy loss function.

5.2.3.4 Inception V3

Inception V3 is developed by the collaboration of Zbigniew Wojna, Christian Szegedy, Sergey Ioffe, Vincent Vanhoucke, and Jonathon Shlens in 2015 (Szegedy et al. 2016).

Inception V3 was the first runner-up for the image classification in the ILSVRC challenge in 2015 by achieving an accuracy of >78.1% accuracy on the ImageNet dataset.

The input image size of the Inception-V3 model is 229 × 229 × 3, i.e., 229 image height × 229 pixel image width × 3 channels (RGB).

We use Keras application Inception-V3 class to create our transfer learning model using "ImageNet" weights along with some modifications as described in Fig. 5.4. We retrained the Inception-V3 model with Adam optimizer along with exponential learning rate decay and binary cross-entropy loss function (Fig. 5.5).

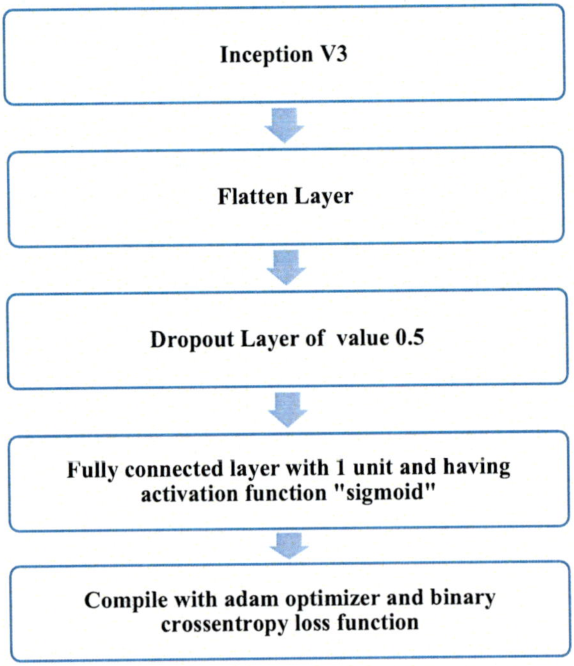

Fig. 5.4 Modified Inception-V3 model

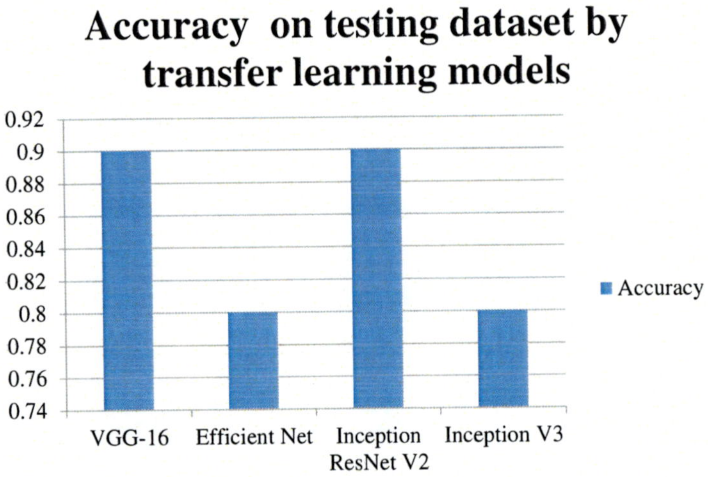

Fig. 5.5 Comparison between accuracies on testing dataset generated by retrained transfer learning models

5.2.4 Training

We train our transfer learning models for 1000 epochs wherein step size per epoch is 3 and regulated with early stopping mechanism which monitors the validation loss every 50 epochs; if the validation loss does not minimize from the last 50 epochs, then training will stop there, but the model will be restored when the validation loss is minimum (Fig. 5.6).

5.2.5 Testing

We tested our trained model with the same testing dataset. For benchmarking of different transfer model, we use different evaluation metrics such as sensitivity, accuracy, specificity, and area under the receiver operating curve. We found out that VGG-16 and Inception-ResNet-V2 perform the same and have the highest accuracies; see Table 5.3.

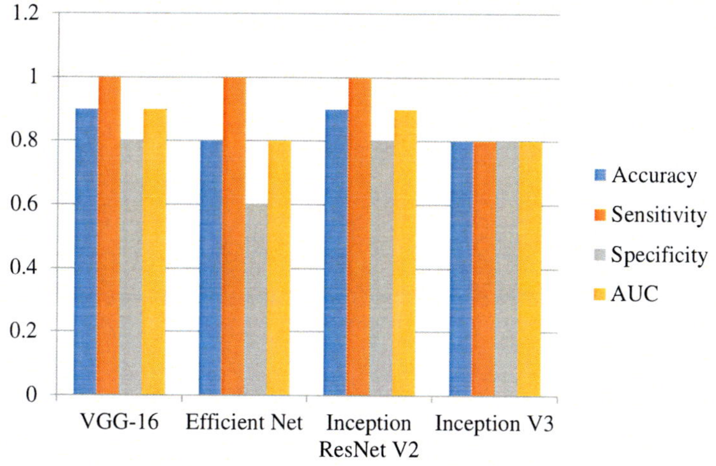

Fig. 5.6 Comparison between various evaluation parameters such as accuracy, sensitivity, specificity, and area under the curve on testing dataset generated by retrained transfer learning models

Table 5.3 *Evaluation parameter results of various models*: we evaluated our transfer learning models using parameters such as accuracy, sensitivity, specificity, and area under the curve

Model	Accuracy	Sensitivity	Specificity	AUC
VGG-16	0.90	1.00	0.80	0.90
EfficientNet	0.80	1.00	0.60	0.80
Inception-ResNet-V2	0.90	1.00	0.80	0.90
Inception V3	0.80	0.80	0.80	0.80

References

Hunt EB, Marin J, Stone PJ (1966) Experiments in induction. Academic Press, Oxford

Kononenko I (2001) Machine learning for medical diagnosis: history, state of the art and perspective. Artif Intell Med 23:89–109. https://doi.org/10.1016/S0933-3657(01)00077-X

LeCun Y, Bengio Y, Hinton G (2015) Deep learning. Nature 521:436–444. https://doi.org/10.1038/nature14539

Michie D, Spiegelhalter DJ, Taylor CC (1994) Machine learning. Neural Stat Classif 13:1–298

Nilsson NJ, Machines L (1965) Foundations of trainable pattern classifying systems. McGraw-Hill, New York

Pan SJ, Yang Q (2010) A survey on transfer learning. IEEE Trans Knowl Data Eng 22:1345–1359

Rosenblatt F (1962) Three layer series coupled perceptrons. In: Principles of neurodynamics: perceptrons and the theory of brain mechanisms. Spartan Books, Washington

Simonyan K, Zisserman A (2018) Very deep convolutional networks for large-scale image recognition Karen. Am J Heal Pharm 75:398–406

Szegedy C, Vanhoucke V, Ioffe S, et al (2016) Rethinking the inception architecture for computer vision. In: Proceedings of the IEEE computer society conference on computer vision and pattern recognition. IEEE Computer Society, pp 2818–2826

Tan M, Le QV (2019) EfficientNet: rethinking model scaling for convolutional neural networks. In: 36th international conference on machine learning ICML 2019, pp 10691–10700

Check for updates

6 Visualization and Prediction of COVID-19 Using AI and ML

Abstract

The global spread of COVID-19, a syndrome of severe respiratory infections, has driven the planet into a global crisis. This would influence each zone, such as the horticultural zone, the agricultural zone, the economic zone, the public transport market, and so on. We published an analysis that identified the effects of the global pandemic using next-generation technologies to see how COVID-19 affected the globe. Prediction is a standard exercise in data science that assists with anomaly identification, objective setting, and strategic planning in administration. We propose a model optimization of interpretable parameters that can clearly be modeled by experts with dataset domain intuition. We focus on international data and conduct complex map simulation of COVID-19's international expansion to date and estimate virus distribution throughout all regions and countries. Detailed overview of both region-wise and state-wise recorded events; forecast of a viral pandemic attack, deaths, and recovered cases; and the degree to which it is spreading globally are included in this chapter.

Keywords

Artificial intelligence · Machine learning · COVID-19

6.1 Introduction

Currently, we are facing new pandemic globally. Day after day, the condition gets serious, since conventional treatment strategies do not prevent it. On November 17, 2019, in Wuhan, the sprawling capital of southern China, the first SARS-CoV-2 pandemic attack was reported. The first case was diagnosed on December 8, 2019, and scientists did not officially accept that there was human-to-human transmission until January 21. COVID-19s symptoms and premonitions are natural, and even now

A. Saxena, S. Chandra, *Artificial Intelligence and Machine Learning in Healthcare*,
https://doi.org/10.1007/978-981-16-0811-7_6

it is difficult to mistake this with another disorder without any corroborative tests. The clinical introduction is that of a serious appearance of respiratory contamination that goes from a moderate common cold-like illness to acute viral pneumonia that induces severe, potentially lethal respiratory distress (Suresh and Jindal 2020). The prevalent virus spreads basically through beads of salivation or discharge from the nose if an affected person hacks or sneezes. The flare-up was announced on January 30, 2020, to be a global public health emergency. Primary risk factors include residence/travel to the area, detailing the network's distribution within 14 days prior to the onset of symptoms, direct contact with an alleged incident, older age, hidden well-being, and malignancy (Suresh 2020).

By March 4, the first signs of frenzy that veils and hydroalcoholic gel would be depleted in quite a while began to bring about open fear about the epidemic. Consequently, the government demanded disposable masks and gloves, along with medically approved sanitizers. Conveying that for individuals displaying signs of disease and for use by health professionals, this type of protective obstruction should be saved separately. Despite this, the cost of hydroalcoholic disinfectant gel was impeded to forestall profiteering. Be that as it might, at present, the supply of covers and hydroalcoholic disinfectant gel is also problematically poor (Ghanchi 2020).

In current situations, systematic mechanisms of testing and recognition (counting touch following), network measures (counting physical separation), improvement of human care programmers, and lighting of the general population, and welfare network can remain a solid focus. To ensure that societies have the flexibility to continue adhering to these measures, it is important to advance mental prosperity for individuals living under physical separation measures. Stringent physical separation steps are especially problematic for society, both economically and psychologically. Therefore, the zeal for characterizing a sound way to cope with deceleration is enormous. In any case, until the incidence of exposure is decreased to an exceedingly low level in a given area, transmission can occur until the maximum of population assurance (Jose et al. 2020) has been achieved.

While focusing on the Indian predicament, the key example of the 2019–2020 coronavirus outbreaks in India was reported from China on January 30, 2020. As of April 14, 2020, the Ministry of Health and Family Welfare announced a total of 10,815 predicaments, 1190 rescued, and 353 registered fatalities in the country. As India's examination rates are among the lowest in the world, experts suggest the number of illnesses could be even higher. On March 24, 2020, Dr. Michal Ryan, the Executive Director of the World Health Organization's Health Emergencies program, declared that India had a "colossal breaking point" to contend with the coronavirus eruption and will have a huge effect on the world's ability to supervise it as the second-most packed country. On March 24, 2020 (https://en.wikipedia.org/wiki/2020_coronavirus_pandemic_in_India), the Prime Minister announced a 21-day shutdown across the nation: the second-most packed region.

A total of 10,815 cases of COVID-19 (including 76 foreign nationals) were registered in 32 states/union territories according to the Ministry of Health and Family Welfare. This includes 1189 restored/released, one moved, and 353 deaths. There is an improvement in the therapeutic separation, monitoring, and home quarantine of contacts from each affirmed case (https://www.who.int/india/emergencies/novel-coronavirus-2019).

6.2 Technology for ML and AI in SARS-CoV-2 Treatment

Near the beginning diagnosis of any disease, whether contagious or noninfectious, is a critical obstacle for premature care to save more lives (Vaishya et al. 2020; Ai et al. 2020). The rapid diagnostic and screening approach seeks to prevent and speed up the eventual diagnosis of the increase of diseases such as COVID-19. By being more cost-effective compared to the traditional approach, the creation of a specialist assists in the current organization of SARS-CoV-2 carrier identification screening and management. Machine learning and artificial intelligence are enhanced by the diagnostic and program procedure of the recognized patient using radio imaging technology similar to X-ray, computed tomography, and blood sample data. The researcher and expert should use scientific images such as CT scans and X-ray as normal methods to boost conventional screening and diagnosis. Sadly, after the high eruption of the SARS-CoV-2 pandemic, the efficiency of such tools is moderate. In this observation, studies (Ardakani et al. 2020) show the capacity of artificial intelligence and machine learning instruments by proposing a novel approach with a fast and true COVID-19 diagnostic mechanism using deep learning. The research reveals the analysis of 108 COVID-19 contaminated patients on 1020 CT images using an expert approach using AI and ML, along with viral pneumonia of 86 patients, that convolution neural network as a tool for radiologists results in 87.11%, 82.21% of precision accuracy, in that order.

With the new approach, automatic identification of COVID-19 based on an AI algorithm (Ozturk et al. 2020), current researchers have created a tool to improve the precision of the diagnosis of COVID-19. With the help of X-ray images, 129 infected patients with 498 without findings and 498 records of cases of pneumonia are included in the built model. Many clusters proved the applicability of the proficient model to quickly and accurately verify the screening process to help radiology.

Researchers have identified 11 main related indices (total protein, bilirubin, basophil, creatine kinase isoenzyme, platelet distribution distance, GLU, calcium, creatinine, lactate dehydrogenase, potassium, and magnesium) after examining 253 Wuhan clinical blood samples, which may help COVID-19 as an important screening discrimination tool for healthcare professionals (Sun et al. 2020; Wu et al. 2020).

Overall, the study provides confirmation of the expert system's implementation; the primary goal was to design rapid diagnosis along with increasing result. In this concept, the detection reduces the progression of the condition and saves some time for the specialist to adapt the next observation and save lives which decreases the cost on medicine. Nevertheless, for most of the analyzed paper, machine learning classification algorithm was used on relevant data. More future multi-domain database algorithms such as clinical, demographic, and mammographic data are then suggested to apply a hybrid classification approach; data has an important feature that can reflect the true identification of patients and the real-world software.

6.3 SARS-Cov-2 Tracing Using AI Technologies

Anticipation of the extent of the illness by contact monitoring is the next crucial phase after an individual is analyzed and confirmed with COVID-19. According to the WHO, the virus spreads primarily by contact transfer from one person to another through sweat and running nose (WHO 2020a). Touch monitoring is an important healthcare method which people use to disrupt the transmission of the disease chain in order to control the spread of COVID-19 (WHO 2020b). The person tracing tool is used to classify and handle public newly showing to a tainted COVID-19 infected people to prevent additional dissemination. Usually the treatment identifies the infected organism with a 14-days go after the following exposure. This method would break the COVID-19 chain of the novel corona virus and reduce the spate by presenting a greater potential for successful helping and controls to minimize the severity of the recent deadly disease. In order to create a digital communication monitoring mechanism with the smartphone application, many infected countries use various technologies such as Global Positioning System (GPS) network-based API, Bluetooth, contact information, social graph, and mobile tracking data. The automated tracing procedure can be real time and easier. These automated technologies are intended to capture data from individual apps that will be processed by artificial intelligence software to track an individual. A study has demonstrated the use of artificial intelligence propelling the pace of contact tracing against COVID-19 diseases (Rorres et al. 2018). After applying the graph theory to data on epidemics of infectious animal diseases, mainly shipping records between each farm, the consequential pictorial properties produced by the planned model can be used to allow contact tracing to be used effectively to improve contact tracing.

Though, presently there are restrictions when resolving situation, anonymity, data management, and still data safety breaches. Many nations, such as Israel, have "passed the emergency law on mobile phone records" to fight this disease (BBC n. d.). In the middle of the worldwide touch tracing applications, some countries apps have broken the confidentiality act and have been reported risky (MIT n.d.) before they do the job properly by supplementing the manual tracing process. Nearly every country, however, has its own touch tracing application, which becomes a public health emergency as the disease continues to spread across the globe. In order to battle this disease, we should have a standard contact tracing programmed to trace any people globally. It is also reported that it is necessary to address any basic question: "Is it compulsory or voluntary?" "Is the initiative transparent or translucent?" "Is the collection of information reduced?" "Will the collected information be demolished as stated?" "Is the host data protected?" "Are there any restrictions or constraints on the use of the data?."

6.4 Forecasting Disease Using ML and AI Technology

A new model, which forecasts and predicts 1 to 7 days to the front of the generally infected COVID-19 individuals in Brazilian states, has been suggested to use the stacking-ensemble with the assist vector regression algorithm on the growing infected COVID-19 cases of this country results, thereby extending the short-term prediction loop to advise the professionals to compensate for the disease (Ribeiro et al. 2020). Using a machine learning classifier named XGBoost on mammographic factor datasets, recent studies have indicated a new method. After applying the algorithm, the experts found that some of the distinctiveness of the 74 experimental and blood test samples (lactic dehydrogenase (LDH), lymphocyte, high-sensitivity C-reactive protein) in estimating and calculating the total number of COVID-19 patients with extreme mortality rates has a median accuracy of 91% (Yan et al. 2020). On the other hand, in identifying the majority of patients in need of intensive medical treatment, the comparatively greater importance of single lactic dehydrogenase tends to be a crucial factor, such that of the degree of LDH involved in various lung illnesses, such as asthma, bronchitis, and pneumonia. The proposed method used the assessment rule to allow patients to be manageable for intensive care and to potentially reduce the rate of transience, in order to easily estimate and forecast infectious people at the greatest risk. Using a deep learning algorithm for the long term, a Canadian-based forecasting model was developed using time series. A key factor in predicting the short-term memory network trajectory was established in the studies, with an end-point prediction of the latest SARS-CoV-2 outbreak in Canada and around the world (Chimmula and Zhang 2020). For this SARS-CoV-2 outbreak in Canada, the proposed end-point model estimate will be around June 2020 (JHU 2020); the prediction was likely to be accurate as newly infected cases dropped rapidly and proved the applicability of the expert approach to predicting and forecasting the next pandemic/epidemic by reestablishing key aspects of veiling. In order to combine the accuracy of the wavelet-based forecasting model with the optimized autoregressive moving average time series model (Chakraborty and Ghosh 2020), the real-time forecasting model was proposed. The model solves the problem by designing short-term SARS-CoV-2 forecasts for various countries as a temporary warning module for each target country to assist healthcare professionals and policy makers.

6.5 Technology of ML and AI in SARS-CoV-2 Medicines and Vaccine

After the beginning of the coronavirus epidemic, scientists and healthcare professionals around the world have been encouraged to develop a potential solution to the development of drugs and vaccines for the SARS-CoV-2 pandemic, and ML/AI technology is an enthralling path. With regard to the likelihood of drug selection for the treatment of infected patients, it is important to provide an urgent review of the existing old, marketable medicines for new SARS-CoV-2 carriers in human beings. Taiwanese researchers are designing a new strategy toward increasing the production of a new drug (Ke et al. 2020). The study revealed eight drugs, i. e., gemcitabine, vismodegib, and clofazimine, using the deep neural network on eighty-year-old drugs with COVID-19 therapeutic potential after two datasets were added to the ML and AI technology-based model (one using 3C-like protease restriction and other data keeping cases of infection with SARS-CoV, SARS-Cov-2, influenza, and human immunodeficiency virus). Additionally, five other drugs, such as salinomycin, homoharringtonine, chloroquine, tilorone, and boceprevir, have also been shown to be operational in the AI laboratory setting.

Researchers from the USA and Korea jointly suggested a novel molecule transformer-drug target interaction model to address the need for an antiviral drug that can cure the COVID-19 virus. The report contrasts AutoDock Vina, a free collaborative screening and molecular docking programmer, to the suggested model, using a deep learning algorithm on COVID-19 3C-like proteins and approved by the FDA, with 3410 new drugs available on the market. The findings found that the best treatment for COVID-19, followed by remdesivir, was a common antiretroviral medication used to treat HIV called atazanavir (Kd of 94.94 nM) (Kd of 113.13 nM). In addition, the findings showed that some drugs for viral proteinase therapy, such as darunavir, ritonavir, and lopinavir, were illuminated. It was also observed that for the medication of COVID-19 human patients, many antiviral compounds such as Kaletra may be used.

An antiviral drug was developed by a group of researchers from the USA to cure the Ebola virus. The study was first made in 2014 (Ekins et al. 2014), beginning with the ML and AI pharmacophore-based statistical study of the small size of in vitro infected carriers of Ebola viruses. The study suggested a widely used amodiaquine and chloroquine complex for the treatment of the malaria virus. In addition, a blend of numerical screening method with docking application and machine learning was introduced after finding a decade of drug development focused on ML and AI technologies to pick supplementary medicine to investigate SARS-CoV-2 (Ekins et al. 2020). Researchers are looking at the successful management of Ebola (Ekins et al. 2020) and the experience of the Zika virus (Ekins et al. 2016), and the same model can also be used to classify COVID-19 drugs and a potential pandemic of the virus.

It was noted that in combination with the docking application, the use of machine software was more effective in forecasting the reusability of an existing old COVID-19 medication and greatly decreasing the amount of a risk factor in creating a more cost-effective drug operation. During this emergency, the use of ML and AI will improve the drug production process by reducing the time slot for the courier to explore an alternative therapy and remedy by depending on a high chance of the efficacy, manageability, and clinical knowledge of the current medicine compound. The finite resources of stable hybrid data and real-life deployment of the programs were the concerns and problems found in this area.

6.6 Analysis and Forecasting

We depend on the daily averages of the three major variables of concern: reported cases, deaths, and recoveries, which are globally aggregated. The Center for Systems Science and Engineering (CSSE) at Johns Hopkins University has retrieved these. These figures are also applied to include the full range of cases spanning the period from January 22, 2020, to March 11, 2020, on an annual basis. They contain both "laboratory-confirmed" and "clinically diagnosed" cases. The significance of recovered cases, which are not as widely mentioned in the media as the cases or deaths identified, is illustrated. In mid-February, the patterns in both registered cases and deaths declined, although all the three data trends increased gradually; in late February and March, a second exponential increase was detected due to a growing number of cases in South Korea, Iran, and Europe. At around the same time, the number of cases that have been collected is gradually increasing.

We accept simple time series forecasting techniques to model recorded COVID-19 incidents. By using models from the exponential smoothing family (Hyndman et al. 2002), we generate predictions. Over several predictive competitions (Makridakis et al. 1982), this family has demonstrated good prediction accuracy and is especially suited for short series. A number of model and seasonal forecasting patterns and their variations can be captured by exponential smoothing techniques. In view of the trends shown in Fig. 6.1, we restrict our focus to trendy and nonseasonal styles. Notice that a sound path is being taken in that we want the development to proceed forever in the future. This methodology opposes other COVID-19 simulation methods, using an S-curve (logistic curve) model that suggests convergence.

Though statistical methods can be used for model selection (such as knowledge parameters that determine the optimum probability of a model while penalizing its complexity), we judgmentally choose a model (Petropoulos et al. 2018) to best represent the essence of the data. Using multiplicative error and multiplicative pattern elements, we chose an exponential smoothing model. While in some situations, taking into account the asymmetric risks involved, an additive trend model offered lower knowledge criterion values, we chose the multiplicative trend model because we believe it is simpler to err in the positive direction (Fig. 6.2).

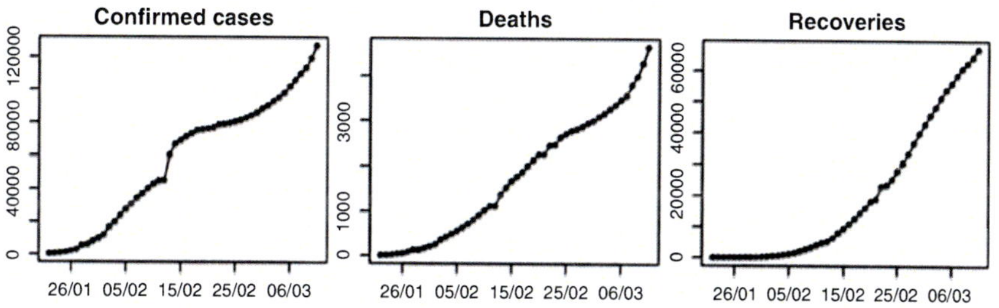

Fig. 6.1 Daily COVID-19 confirmed, death, and recovered cases

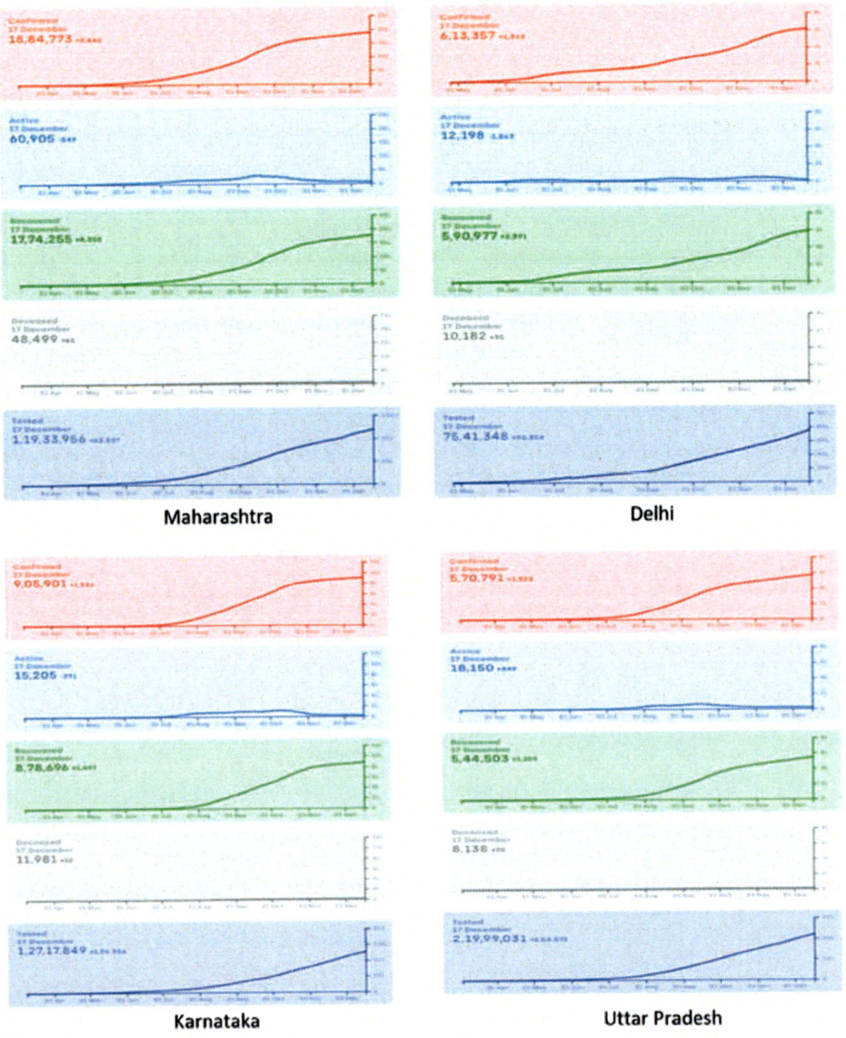

Fig. 6.2 Highly affected regions for COVID-19 confirmed, active, recovered, and tested cases in India

6.6.1 Predictions on the First Round

We first started at the end of January 30, 2020, and had only ten actual data points in our hands. We have to make use of the exponential smoothing model of a multiplicative pattern. The forecasts were made at the end of January 30. For the 10-day-ahead events reported, the mean estimate (point forecast) was 209,000, with 91% estimation periods ranging from approximately 39 to 535,000 instances. The actual cases confirmed were just under 42,000 on February 11, 2020. We found a significant predicted loss equal to 166,000 instances from the normal calculation (a cumulative percentage error of 389%), with the figures being highly positive. However, the individual cases fell behind the prediction intervals.

6.6.2 Predictions on the Second Round

At the end of February 10, 2020, the chronological sum of our data was then broadened to include incidents. We made the 10-day-ahead predictions yet again. For the time between February 12, 2020, and February 21, 2020, it should be noted that the average estimate is closely followed by the real values. For February 21, 2020, the estimated error is 5.9 thousand instances. Notwithstanding the adjustment made on February 14, 2020, as opposed to laboratory-proven instances, "clinically identified" instances are now included only in terms of how reported cases are registered. One interesting observation is that this more robust forecast resulted in a substantial decrease in the slope's steepness relative to the previous 10-day duration forecast. Another point is that we have already overestimated the number of confirmed cases at the end of February 21, 2020. Lastly, all the real values were way above the predicted range of intervals.

6.6.3 Predictions on the Third Round

We generated a third set of predictions and prediction dates using the data up to February 21, 2020. The mean estimate for 11 days in advance was 85,000 instances. Again, the slope of the forecasts was lower than that of the previous two forecast sets, which confirmed a steady decline in the number of cases registered. We also noted a significant decrease in the corresponding projection volatility relative to our previous predictions, with the prediction periods becoming marginally tighter. For 91% of projected periods, the worst-case situation was almost 700,000 examples, which is half according to the current round of forecasts. The real cases confirmed were 87,000 at the end of March 02, 2020. We reported an error of 56,000 instances at the conclusion of this third round of calculations.

6.6.4 Predictions on the Fourth Round

The cumulative estimate for March 12, 2020, was 113,000 confirmed occurrences, with a comparable volatility rate to the previous round: at the end of March 12, 2020, there was a 6% probability that they would reach 614,000. The actual confirmed events reported at the end of this time are about 128,000. At the end of the last period, the absolute forecast error was 15.5 K, higher in comparison to the previous forecast series but still well within the forecast intervals. We were continuously under-forecasting the real events for the second round in a row. This was attributed to an extraordinary surge in the number of new cases reported, mostly in Europe, Iran, and the USA, with South Korea being able to decrease the number of new cases each day significantly.

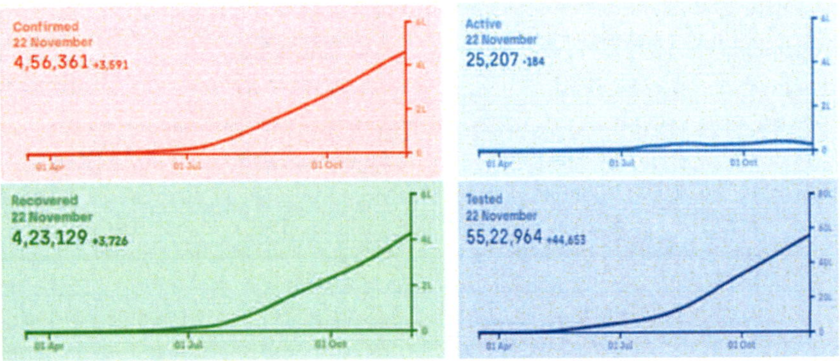

Fig. 6.3 COVID-19 confirmed, active, recovered, and tested cases in India

6.6.5 Predictions on the Fifth Round

We produced a final set of forecasts and prediction intervals until March 12, 2020, using the most recent proof. Notice that we have calculated 3 degrees of uncertainty. Compared to the last two rounds, the trend in our projections is even higher: for this round, we predict 84,000 new incidents. The associated levels of volatility are much higher: there is a 26% chance that by the end of March 22, 2020, the overall registered cases will reach 414,000 and a 6% probability that by the end of March 22, 2020, they will touch 1.18 million (Fig. 6.3).

By segregating the reported confirmed cases into two types—cases within Mainland China and cases somewhere else—we have tried to build forecasts since the distinctions between these two classes are different. We developed exponential smoothing models separately and then summarized the forecasts. Using this method, the average estimation is equal to that of all the data considered together, we remember. The evidence of variation splitting, however, is estimated to be considerably smaller, as recorded cases outside Mainland China are only likely to have increased dramatically recently.

6.7 Methods Used in Predicting COVID-19

6.7.1 Recurrent Neural Networks (RNN)

Deep learning speculates that a deep sequential or hierarchical model is more effective than shallow models (Bengio 2009) in classification or regression functions. There are implied states distributed over time in recurrent neural networks, and this helps them to retain a lot of previous knowledge. Due to their ability to handle variable length sequential data, they are most widely used in forecasting applications (Graves 2013). There is a major drawback to recurrent neural networks that they do not respond to the gradient disappearance or gradient explosion problem and can only store short-term memory and have only hidden layer activation functions in the previous step (Hochreiter and Schmidhuber 1997).

6.7.2 Long Short-Term Memory (LSTM) and Its Variants

It is known that LSTMs are among the most efficient solutions for prediction operations, and based on the different highlighted features present in the dataset, they forecast future predictions. With LSTMs, knowledge travels through elements known as cell states. LSTMs may recall or miss details correctly. The data obtained over progressive stretches of time is known to be the data from time series, and LSTMs are typically used as a rigorous means of calculating these data values. The model converts the previous veiled state to the appropriate stage of the arrangement in this style of architecture. Long short-term memory cells (Hochreiter and Schmidhuber 1997) are used for long-term memory retrieval RNNs, while RNNs can retain only a small amount of information. The problems of the gradient disappearing and the bursting gradient (Bengio et al. 1994) plaguing RNN are resolved by LSTMs. LSTM cells are similar to RNN, with memory blocks replaceable by hidden modules.

6.7.3 Deep LSTM/Stacked LSTM

The regular LSTM extension we have defined above is stacked LSTM (Graves et al. 2013), also known as Deep LSTM. There are several hidden layers and several memory cells on the stacked LSTM. The depth of the neural network is improved by the stacking of multiple layers, where each layer has some information and transfers it to the next. The top LSTM layer supplies the previous layer with sequence information and so on. For each time step, it produces a different output instead of a single output for all time steps.

6.7.4 Bidirectional LSTM (Bi-LSTM)

Inputs are processed by traditional RNNs in only one direction and ignore the possible knowledge that they provide. By following the bidirectional topology of LSTM (Schuster and Paliwal 1997), this issue is solved. By keeping both past and future information into account, bidirectional LSTM (Bi-LSTM) excludes absolute temporal time information. Periodic secret RNN neurons are separated into forward and backward states in which forward state neurons are not bound to backward states and vice versa. The design without the backward states is identical to the normal unidirectional RNN. There is no need to provide extra time delays as used for this process in standard RNN.

6.8 Conclusion

A comparative discussion of reported infections, recovered cases, and mortality status across numerous countries on the globe is seen in the COVID-19 pandemic infection prediction study using machine learning and AI. When we approach the state of sickness, the lack of appropriate social distance and personal hygiene plays an important part in adding to the prevalent community. Effective management may track the progress of the illness to a limited degree using symptomatic care and quarantine equipment. In the future, there might be questions for human life if the condition grows worse. To approximate the number of positive cases of COVID-19, we have also suggested deep learning models in Indian states. An exploratory data analysis on the rise in the number of positive cases in India has been undertaken. States are graded state wise into medium, moderate, and severe zones based on the number of cases and the periodic development rate for realistic shutdown measures, as opposed to shutting the entire country down, which may trigger socioeconomic problems. These predictions will be helpful for state and national government leaders, consultants, and planners in order to prepare hospitals and coordinate medical services accordingly. Many countries are already prepared to follow the proposed model and defense strategy.

References

Ai T, Yang Z, Hou H, Zhan C, Chen C, Lv W, Tao Q, Sun Z, Xia L (2020) Correlation of chest CT and RT-PCR testing in coronavirus disease 2019 (COVID-19) in China: a report of 1014 cases. Radiology 296:32–40. https://doi.org/10.1148/radiol.2020200642

Ardakani AA, Kanafi AR, Acharya UR, Khadem N, Mohammadi A (2020) Application of deep learning technique to manage COVID-19 in routine clinical practice using CT images: results of 10 convolutional neural networks. Comput Biol Med 103795(2020):121. https://doi.org/10.1016/j.compbiomed.2020.103795

BBC (n.d.) Coronavirus: Israel enables emergency spy powers. https://www.bbc.com/news/technology-51930681. Accessed 3 Jun 2020

Bengio Y (2009) Learning deep architectures for ai. Found Trends Mach Learn 2(1):1–127

Bengio Y, Simard P, Frasconi P (1994) Learning long-term dependencies with gradient descent is difficult. IEEE Trans Neural Netw 5(2):157–166

Chakraborty T, Ghosh I (2020) Real-time forecasts and risk assessment of novel coronavirus (COVID-19) cases: a data-driven analysis. Chaos Solitons Fract 135:109850. https://doi.org/10.1016/j.chaos.2020.109850

Chimmula VKR, Zhang L (2020) Time series forecasting of COVID-19 transmission in Canada using LSTM networks. Chaos Solitons Fract 135:109864. https://doi.org/10.1016/j.chaos.2020.109864

Ekins S, Freundlich J, Coffee M (2014) A common feature pharmacophore for FDA-approved drugs inhibiting the Ebola virus. F1000Research 3:277

Ekins S, Mietchen D, Coffee M, Stratton TP, Freundlich JS, Freitas-Junior L, Muratov E, Siqueira-Neto J, Williams AJ, Andrade C (2016) Open drug discovery for the Zika virus. F1000 Res 5:150. https://doi.org/10.12688/f1000research.8013.1

Ekins S, Mottin M, Ramos PRPS, Sousa BKP, Neves BJ, Foil DH, Zorn KM, Braga RC, Coffee M, Southan C, Puhl CA, Andrade CH (2020) Déjà vu: stimulating open drug discovery for SARS-CoV-2. Drug Discov Today 25:928–941. https://doi.org/10.1016/j.drudis.2020.03.019

Ghanchi A (2020) Adaptation of the National Plan for the prevention and fight against pandemic influenza to the 2020 COVID-19 epidemic in France. Disaster Med Public Health Prep 7:1–3. https://doi.org/10.1017/dmp.2020.825

Graves A (2013) Generating sequences with recurrent neural networks. arXiv preprintarXiv:130808502013

Graves A, Mohamed AR, Hinton G (2013) Speech recognition with deep recurrent neural networks. In: 2013 IEEE international conference on acoustics, speech and signal processing. IEEE, pp. 6645–6649

Hochreiter S, Schmidhuber J (1997) Long short-term memory. Neural Comput 9(8):1735–1780

Hyndman RJ, Koehler AB, Snyder RD, Grose S (2002) A state space framework for automatic forecasting using exponential smoothing methods. Int J Forecast 18(3):439–454

JHU (John Hopkins University) (2020) COVID-19 Dashboard by the Center for Systems Science and Engineering (CSSE) at Johns Hopkins University (JHU). https://www.coronavirus.jhu.edu/map.html. Accessed 9 Jun 2020

Jose J, Yuvaraj E, Aswin S, Suresh A (2020) Development of worldwide tsunami hazard map for evacuation planning and rescue operations. Preprints 2020:2020040370. https://doi.org/10.20944/preprints202004.0370.v1

Ke Y-Y, Peng T-T, Yeh T-K, Huang W-Z, Chang S-E, Wu S-H, Hung H-C, Hsu T-A, Lee S-J, Song J-S, Lin W-H, Chiang T-J, Lin J-H, Sytwu H-K, Chen C-T (2020) Artificial intelligence approach fighting COVID-19 with repurposing drugs. Biom J 43:355–362. https://doi.org/10.1016/j.bj.2020.05.001

Makridakis S, Andersen A, Carbone R, Fildes R, Hibon M, Lewandowski R et al (1982) The accuracy of extrapolation (time series) methods: results of a forecasting competition. J Forecast 1(2):111–153

MIT (n.d.) Covid tracing tracker—a flood of coronavirus apps are tracking us. Now it's time to keep track of them. https://www.technologyreview.com/2020/05/07/1000961/launching-mittr-covid-tracing-tracker/. Accessed 5 Jun 2020

Ozturk T, Talo M, Yildirim EA, Baloglu UB, Yildirim O, Rajendra Acharya U (2020) Automated detection of COVID-19 cases using deep neural networks with X-ray images. Comput Biol Med 121:103792. https://doi.org/10.1016/j.compbiomed.2020.103792.

Petropoulos F, Kourentzes N, Nikolopoulos K, Siemsen E (2018) Judgmental selection of forecasting models. J Oper Manag 60:34–46

Ribeiro MHDM, da Silva RG, Mariani VC, Coelho LDS (2020) Short-term forecasting COVID-19 cumulative confirmed cases: perspectives for Brazil. Chaos Solitons Fract 135:109853. https://doi.org/10.1016/j.chaos.2020.109853

Rorres C, Romano M, Miller JA, Mossey JM, Grubesic TH, Zellner DE, Smith G (2018) Contact tracing for the control of infectious disease epidemics: chronic wasting disease in deer farms. Epidemics 23:71–75. https://doi.org/10.1016/j.epidem.2017.12.006

Schuster M, Paliwal KK (1997) Bidirectional recurrent neural networks. IEEE Trans Signal Process 45(11):2673–2681

Sun L, Liu G, Song F, Shi N, Liu F, Li S, Li P, Zhang W, Jiang X, Zhang Y, Sun L, Chen X, Shi Y (2020) Combination of four clinical indicators predicts the severe/critical symptom of patients infected COVID-19. J Clin Virol 128:104431. https://doi.org/10.1016/j.jcv.2020.104431.

Suresh A (2020) Mystery over the Haze during 1st week of November 2019 in Delhi-NCR. Preprints 2020:2020040156. https://doi.org/10.20944/preprints202004.0156.v1

Suresh A, Jindal T (2020) Phthalate toxicity. https://doi.org/10.20944/PREPRINTS202004.0209.V1

Vaishya R, Javaid M, Khan IH, Haleem A (2020) Artificial intelligence (AI) applications for COVID-19 pandemic. Diabetes Metab Syndr 14(4):337–339. https://doi.org/10.1016/j.dsx.2020.04.012.

WHO (World Health Organization) (2020a) Health topic, coronavirus disease overview. https://www.who.int/health-topics/coronavirus#tab=tab_1. Accessed 29 May 2020

WHO (World Health Organization) (2020b) Contact tracing in the context of COVID-19. https://www.who.int/publications-detail/contact-tracing-in-the-context-of-covid-19. Accessed 29 May 2020

Wu J, Zhang P, Zhang L, Meng W, Li J, Tong C, Li Y, Cai J, Yang Z, Zhu J, Zhao M, Huang H, Xie X, Li S (2020) Rapid and accurate identification of COVID-19 infection through machine learning based on clinical available blood test results. medRxiv. https://doi.org/10.1101/2020.04.02.20051136

Yan L, Zhang H-T, Goncalves J, Xiao Y, Wang M, Guo Y, Sun C, Tang X, Jing L, Zhang M, Huang X, Xiao Y, Cao H, Chen Y, Ren T, Wang F, Xiao Y, Huang S, Tan X, Huang N, Jiao B, Cheng C, Zhang Y, Luo A, Mombaerts L, Jin J, Cao Z, Li S, Xu H, Yuan Y (2020) An interpretable mortality prediction model for COVID-19 patients. Nat Mach Intell 2:283–288. https://doi.org/10.1038/s42256-020-0180-7

Machine Learning Approaches in Detection and Diagnosis of COVID-19

7

Abstract

Novel coronavirus disease (COVID-19) has hit the world in December, 2019, with the first case being identified in Wuhan, China. Since then, international health agencies are making serious efforts to manage the pandemic, exploring every aspect of therapy development with a special attention on investigating smart diagnostic tools for rapid and selective detection of COVID-19. Detection of the disease is mainly through reverse transcription-polymerase chain reaction (RT-PCR) test, which is complex, expensive, and time-consuming, making it difficult to scale-up for mass testing. Hence, there is a need for parallel diagnostic testing procedures that are fast, accurate, and reliable. In many recent studies, it has been shown that COVID-19 disease clearly exhibits distinct infection patterns in the lung distinguishable from other pneumonia-related diseases. Machine learning and artificial intelligence are well-established methods in image analysis, making them suitable for the analysis of computerized tomography (CT) chest scans and X-ray images. This provides a novel class of testing that is noninvasive and can help in point-of-care testing by the use of portable CXR machines. AI-based medical imaging can help in quickly and accurately labeling specific abnormal structures, without omission of even small lesions, making them suitable for the analysis of chest CT scans and X-ray images. This would alleviate the growing burden on radiologists and assist them in making accurate diagnosis. In this chapter, we present an overview of the state-of-the-art deep learning architectures in the detection of COVID-19 by analysis of chest CT scans and X-ray images.

Keywords

COVID-19 · Deep learning · Chest X-rays · Computerized tomography scans · Convolutional neural network · ResNet · DenseNet · Inception · Xception

A. Saxena, S. Chandra, *Artificial Intelligence and Machine Learning in Healthcare*,
https://doi.org/10.1007/978-981-16-0811-7_7

7.1 Introduction

The Coronavirus disease (COVID-19) is a pulmonary infection triggered by a newly discovered severe acute respiratory syndrome coronavirus 2 (SARS-CoV-2). First diagnosed in the city of Wuhan in China in December 2019, it soon became pandemic, affecting lives and economy globally. Within 6 months, there are over 25 million cases and more than 6 lakh deaths globally, which is a highly underestimated figure of the actual number of cases/deaths. The disease manifestation in infected individuals ranges from asymptomatic infection to mild and moderate symptoms to critical respiratory illness, requiring ventilators and hospitalizations. High proportion of asymptomatic or mild infections, ~80–85%, has made it difficult to assess the true extent of the spread of the virus and its infection-fatality ratio. Since there are no vaccines or treatments available till date for mitigating the spread of the disease, early detection, isolating the infected individual, contact tracing, and quarantining those in contact with the infected individuals are the methods adopted worldwide to contain the spread of the disease. Medical resources being limited in most regions, faster diagnosis, and early detection of high-risk COVID-19 patients are desirable for prevention and optimization of the resources.

Diagnosis involves real-time reverse transcription-polymerase chain reaction (RT-PCR) test for the presence of virus in oral/nasal specimens. Though considered the gold standard in COVID-19 detection, false-negative rate of RT-PCR is shockingly high, ~100–67%, within the first 5 days of exposure and lowest on the eighth day of exposure (20%), increasing again to 66% by day 21 (Kucirka et al. 2020). Thus, differential response to SARS-CoV-2 in different people, day of sample collection after exposure, and incubation period of the virus are the major factors affecting the final outcome of the RT-PCR. Hence, RT-PCR result alone cannot be used to rule out COVID-19 infection. Another popular diagnostic test is a simple blood test (serological test) that measures the presence of antibodies that our immune system makes to defend against SARS-CoV-2 infection, which our bodies continue to make even after the virus is eliminated, irrespective of whether the individual had mild, severe, or no symptoms. From around 2 days to 3 weeks after infection, an individual would start producing IgM antibodies. After a few days, its production declines, and the body starts making IgG antibodies. However, the serological tests are not very accurate (with as low as 30% accuracy). Quicker and potentially portable methods are underdevelopment, for example, reverse transcriptase loop-mediated isothermal amplification (RT-LAMP) (Thi et al. 2020) or a gene-editing method called CRISPR (Mojica et al. 2009; Cong et al. 2013).

Shortage of testing kits, even in developed countries, has led to increase in efforts for finding alternate solutions with high sensitivity. Alternate nonmolecular techniques have been used as initial screening method: analysis of chest radiography images, viz., chest X-ray (CXR), and computer tomography (CT) scans for identifying COVID-like infections in the lungs. A chest radiograph, called a chest X-ray, is routinely used in medical imaging, prescribed to diagnose conditions affecting the chest, its contents, and neighboring structures. The X-ray films of pneumonia, generally caused by bacteria, viruses, mycoplasma, and fungi, are characterized by features such as airspace opacity, lobar consolidation, or interstitial opacities. Because chest X-ray provides a noninvasive, fast, and easy test, it is particularly useful in emergency diagnosis and treatment. In computer-aided medical image analysis for diagnosis, CXR image classification is an active research area. Chest computed tomography, commonly known as CT scan, is another fast, noninvasive, and accurate medical imaging diagnostic test of the chest that is more sensitive than traditional X-ray images. In this imaging technique, multiple X-ray measurements of the lung are taken at different angles, called slices. A computer then combines these slices to generate a 3D model that help to show the size, shape, and position of lungs and structures in the chest. Though CT scans provide more accurate diagnosis, the availability of portable X-ray machines and cost makes them the first line of examination.

Pneumonia detection using CXR and CT image analysis has been the standard practice globally for many years. Not only pneumonia but also various other pulmonary diseases, including SARS and MERS detection, are also done using analysis of CXR images by expert radiologists. In fact, the early detection of this new disease, COVID-19, was identified by using CT scans. When group of patients with typical symptoms of respiratory infection were admitted in Hubei, China, CT scans of these patients showed varied opacities compared to images of normal people and were initially diagnosed to have common pneumonia. Multiplex RT-PCR assay of previously known pathogen panels resulted in negative results, suggesting the infection to be caused by unknown species. Fang et al. carried out one of the first studies to assess the reliability of CT scans over RT-PCR, and the flowchart of the study in Fig. 7.1 clearly highlights the importance of this alternate approach (Fang et al. 2020). Their findings resulted in higher sensitivity (98%) using CT image analysis compared to RT-PCR (71%) with $p < 0.001$, supporting the use of chest CT to screen COVID-19 patients.

Another parallel study on a larger cohort of 1014 subjects in Wuhan, China, during the month of January, 2020, has assessed the reliability of CT scans as a diagnostic tool for COVID-19 compared to RT-PCR. It was observed that 601 of 1014 patients (59%) exhibited positive results with RT-PCR, while 888 of 1014 patients (88%) resulted in positive results, using chest CT scans. Of 413 patients that gave negative RT-PCR, 308 patients (48%) exhibited positive results with CT scans, indicating the sensitivity of it as a diagnostic tool for COVID-19 (Ai et al. 2020). These early studies clearly indicate the use of CXR or CT imaging as a primary tool for testing before RT-PCR or along with RT-PCR for identifying the priority in admissions of patients into hospitals and ICUs. Numerous studies have further confirmed their use in the clinical setting.

In many recent studies, it has been shown that the COVID-19 disease indicates distinct infection patterns in the lungs, which are distinguishable from other pneumonia-related diseases. The typical features of COVID-19 disease in chest CT images include changes in the lung, such as consolidation, i.e., accumulation of fluid and/or tissue in pulmonary alveoli, ground-glass opacity, and nodular shadowing, along with the periphery and lower areas of lungs. Vascular dilations

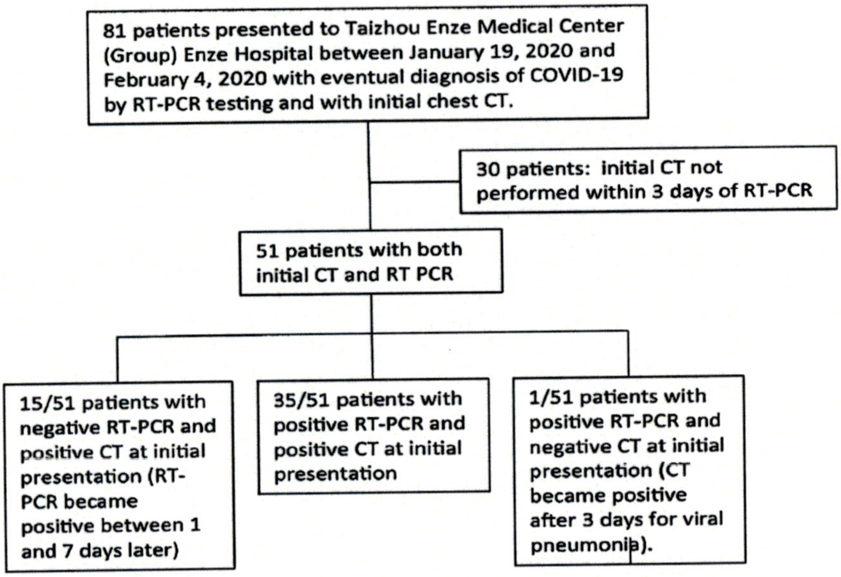

Fig. 7.1 Flowchart of the study by Fang et al. to assess the performance of CT scans for the detection of COVID-19 comparison to RT-PCR (reproduced from (Fang et al. 2020))

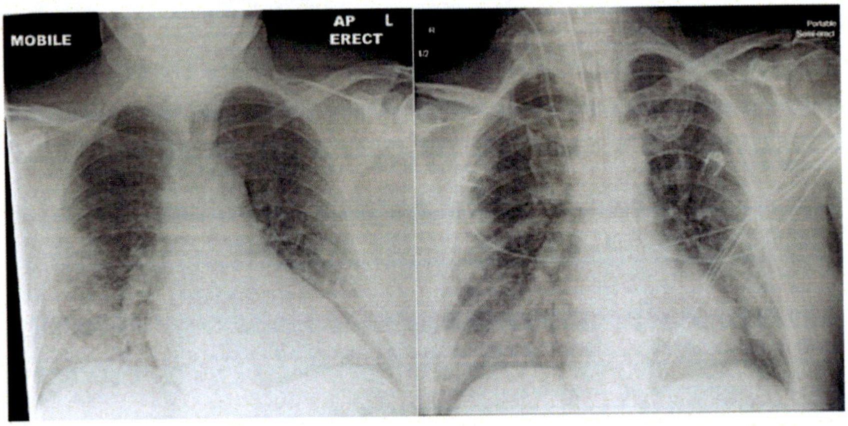

Fig. 7.2 Chest X-ray image on day 3 of a COVID-19 patient (left) clearly indicates right mid and lower zone consolidation; on day 9 (right) is seen worsening oxygenation with diffuse patchy airspace consolidation in the mid and lower zones. (Case courtesy of Dr. Derek Smith, Radiopaedia.org, rID: 75249)

and crazy paving (thickened interlobular lines) are other patterns that are seen in CT images of the COVID-19 patients (Ng et al. 2020; Awulachew et al. 2020). The characteristics of the disease is clearly seen in X-ray (Fig. 7.2) and CT scan (Fig. 7.3) images of the chest in COVID-19 patient that help in distinguishing coronavirus

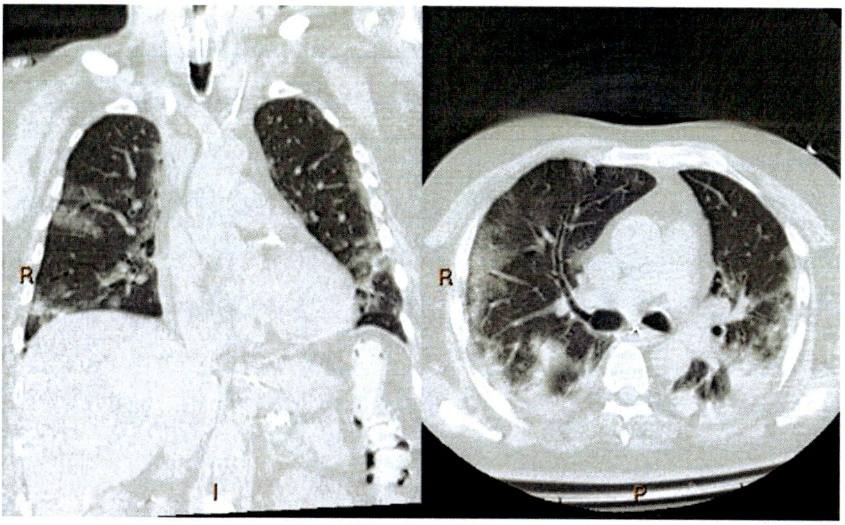

Fig. 7.3 CT scan image performed to assess the degree of lung injury of the patient in Fig. 7.2 on day 13 (left coronal lung window, right axial lung window). Multifocal regions of consolidation and ground-glass opacifications with peripheral and basal predominance. (Case courtesy of Dr. Derek Smith, Radiopaedia.org, rID: 75249)

infection from other pulmonary infections, seen as white patches in CXR and CT images. Thus, CXR and CT imaging provides a noninvasive diagnostic testing that can help in point-of-care testing by the use of portable CXR machines. However, chest radiography image analysis requires the expertise of radiologists, which may create a bottleneck in the decision-making process during a pandemic situation. The use of artificial intelligence (AI) models for the diagnosis of COVID-19 using thoracic images helps in triaging patients and reducing turnaround time, by automating the decision-making process (Chandra and Verma 2020; Varshni et al. 2019). This has led to the deployment of computer-aided systems using machine learning (ML) methods in various hospitals globally to help medical staff in faster triaging of patients.

Deep learning (DL) methods have proven to be the cutting-edge image analysis tools in most fields. Because of their ability to capture patterns in input images, DL methods find application in varied tasks, such as face recognition (Mehendale 2020; Nagpal et al. 2019), object identification (Pérez-Hernández et al. 2020; Ren et al. 2016), applications in natural language processing (Guo et al. 2020), and medical image analysis (Smailagic et al. 2020; Spanhol et al. 2016). Convolutional neural network (CNN) is one of the popular architectures of DL models applied in image analysis. Various CNN architectures, such as ResNet (He et al. 2015), DenseNet (Huang et al. 2018), Inception (Szegedy et al. 2015), etc., have been proposed for various tasks based on the data type and application. Here, an overview of some CNN-based studies is proposed for the diagnostic and prognostic analysis of chest radiography images of COVID-19 suspects, including domain knowledge aware models. The performance of these models is shown to be comparable to that of human experts. Limitations and drawbacks of these models, such as lack of sufficient data for training the models, unavailability of annotated data, difficulty in interpreting the results, etc., and methods to overcome these are discussed.

7.2 Review of ML Approaches in Detection of Pneumonia in General

Pneumonia is the swelling of air sacs in the lungs due to various reasons, including viruses, bacteria, fungi, etc., and may result in minor to severe illness in people of all ages. Common symptoms of pneumonia include cough, fever, shortness of breath, etc., and diagnosis mainly involves chest X-ray, blood culture, sputum culture, CT scan, etc. Treatment of pneumonia is based on the type of infection: bacterial, viral, or fungal. Various viral infections from common flu (influenza) to the deadly ones, including SARS, MERS, etc., can lead to pneumonia, and death in most such cases is due to respiratory failure. The CXR and CT scans are the standard practices for detecting pneumonia, but in pandemic-like situations, the imaging departments in hospitals may get overwhelmed by the huge number of cases, as analyzing radiography images needs expertise and time. Thus, there clearly is a need for automatic analysis of these images, and AI/ML-based technologies are well-suited for such tasks. Lung infections typically look as opaque areas in the images, which can be unclear and difficult to distinguish between various lung abnormalities, like pneumothorax, pleural effusion, pneumonia, pulmonary tuberculosis, lung scarring, etc., posing a challenge even to radiologists. ML-based systems can assist radiologists in arriving at the correct decisions as shown by various studies in detecting pulmonary diseases, like diagnosing pulmonary tuberculosis (Lakhani and Sundaram 2017), classification of lung nodules for lung cancer detection (Hua et al. 2015), and detection of various other abnormalities from radiography images (Islam et al. 2017). By identifying eight statistical features of the segmented lung areas, Chandra and Verma (2020) were able to classify CXR images into pneumonia and normal. Deep learning (DL) methods have proved the best among other ML approaches in the classification tasks for image data (Antin et al. 2017; Sedik et al. 2020). In fact, application of ML techniques in medical image diagnosis has proved its ability in reaching human-level expertise now (Rajpurkar et al. 2017; Jin et al. 2020).

7.3 Application of Deep Learning Approaches in COVID-19 Detection

Pandemic situations like COVID-19 pose several limitations to human interventions in handling the situation. All the complications from the risk of transmission of disease to healthcare professionals to delay in detection and isolation of patients can be better managed using an appropriate application of technology. The earlier AI=/ ML-based studies on pulmonary diseases indicate that the diagnosis of COVID-19 can be quicker and consistent with DL techniques, which provide cutting-edge technology in image analysis applications. Thousands of radiography images have been generated during the past few months since the outbreak of COVID-19. These images are being be used to train DL models, for assessing the risk of patients developing pneumonia from coronavirus infection, and to screen the status of patient's lungs over the course of infection, by carrying out serial imaging of the chest for comparison. Thus, through the preliminary AI screening and diagnosis, not only the higher diagnosis quality can be guaranteed, by reducing the omission of small lesions, but also bring a significant cost reduction and better management of hospital resources.

A brief overview of a few recently proposed DL models for successfully classifying COVID-19 images from that of other pneumonia and normal images is provided. Convolutional neural networks (CNN) recently gained lot of attention among other DL models in the diagnosis of COVID-19. The review is organized as follows. In Sect. 7.3.1, the basic framework of the DL models for the detection of COVID-19 is discussed. Section 7.3.2 discusses one of the common challenges faced by all the models, i.e., lack of available data leading to class imbalance and transfer learning approaches adopted to overcome the challenge. In the last section, the methods used for the interpretation of the results, called explainable learning models, and visualizing the features extracted by these models are briefly described.

7.3.1 Deep Learning Model Frameworks

Convolutional neural networks (ConvNet/CNN) are deep learning methods that typically take an image as input and has an architecture meant to support the image dimensions. The neurons in the different layers of a CNN are arranged in the three dimensions, height, width, and depth, similar to the connectivity of neurons in the brain. Neurons in each layer are connected to a small region of the previous layer. Typically, a CNN consists of three layers, viz., convolutional layer, pooling layer, and a dense (fully connected) layer, as shown in Fig. 7.4. These layers are stacked to form a ConvNet architecture, and their function is briefly described below.

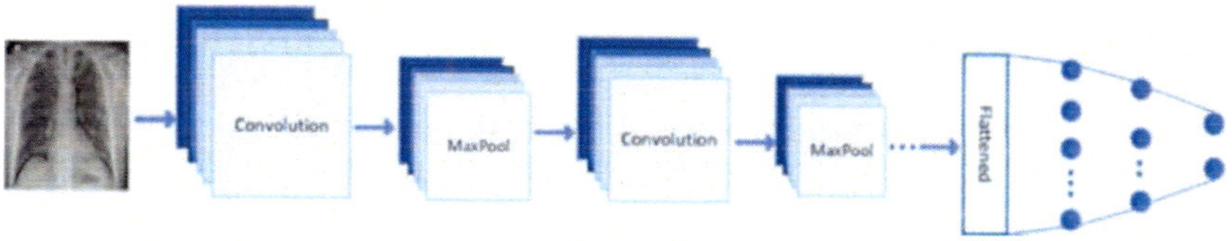

Fig. 7.4 Typical convolutional network framework for classifying COVID-19 cases, which takes as input CXR images and passes through a series of convolution, pooling, and dense layers and uses a softmax function to classify an image as COVID-19 infected with probabilistic values between 0 and 1

Convolutional Layer In a convolutional layer, neurons are arranged in three dimensions, namely, height, width, and depth, and each layer transforms the input volume to an output volume of activations. In each layer, the neurons are connected only to the local regions of the input, and dot product is computed between the weights and the input from only those local regions. Each of these is considered a filter, and these filters shifts through the entire image in a number of strides. Multiple such filters can be applied to the input volume in a layer. The convolutional layer detects the features from the small localized regions, common across the input data, and generates feature maps. These feature maps are fed to an activation function, such as tanh, Sigmoid, ReLU, Leaky ReLU, etc., introducing nonlinearity in the output of the convolution layer, and yield a transformed output.

Pooling Layer A pooling layer typically does down-sampling along the spatial dimensions. That is, the size of the input is reduced resulting in a reduction in the parameters of the network. Average pooling, max pooling, etc. are some of the pooling functions applied. A pooling layer of size 2x2 down-samples every slice depth-wise along both height and width dimensions, by taking the average (average pooling) or max value (max pooling) of the 2x2 regions; thereby, depth remains unchanged, while width and height dimensions get reduced.

Dense Layer Also known as the fully connected layer, it computes the final class scores of an input; hence, it results in a volume of size 1x1xN, where N is the number of classes. Each neuron in this layer connects all the outputs from the previous volume. The features extracted from the preceding layers are analyzed globally in this layer, and a nonlinear combination of these features is subjected to a classifier. Based on how strongly the features map to a particular class, a score is generated by the activation function. Other than these layers, optional layers, such as batch normalization layer, dropout layer, etc., are added to address the problems of slow convergence and overfitting, respectively.

Convolutional layer and fully connected layers have weight parameters associated with them, whereas pooling layer does not. The architecture of a CNN helps in reducing the number of parameters required for learning a model compared to a regular neural network, as the number of inputs from the images is very high and computing dot products of all the weights and inputs in a number of fully connected neural network results in a huge number of parameters. There are several popular architectures of CNNs, LeNet, AlexNet, GoogLeNet, Inception (Das et al. 2020), VGG (Brunese et al. 2020), ResNet, etc. Of these, ResNet models are the most widely applied architecture in the COVID-19 analysis applications. LeNet was first of its kind in the family of CNNs, which had five alternating convolution and pooling layers, followed by two fully connected (dense) layers. AlexNet improved upon the architecture of LeNet by adding a few more layers to it and making additional parameter modifications, such as using large-sized filters, skipping a few transformational units during training, etc. VGG was introduced later with an increased depth with 19 layers but reducing the size of the filters compared to the previous versions. But the depth of the network introduced an overhead of training 138 million parameters, which makes it unaffordable for low resource system applications. GoogLeNet introduced the concept of inception blocks, where the traditional convolution layers are replaced with smaller blocks and have filters of different sizes to capture patterns at different scales. The architecture of GoogLeNet applied various parameter optimizations, such as discarding redundant feature maps, using global average pooling, etc., which helped in limiting the number of parameters to four million. There are many other variants of CNNs, which were introduced by making changes to the architecture (not covered here). A detailed review is given in the study by Khan et al. (2020a). Some of the most popular architectures applied in the detection of COVID-19 reviewed in this chapter are given in Table 7.1.

7.3.1.1 ResNet Models

Various DL methods proposed for the detection of COVID-19 have used a residual network (ResNet) architecture, namely, COVID-Net (Wang and Wong 2020), CoroNet (Khobahi et al. 2020), CovNet (Li et al. 2020), and models proposed by Jin et al. (2020) and Gozes et al. (2020), to name a few. ResNet models are currently the default choice for implementing convolutional networks (Bressem et al. 2020; Waleed Salehi et al. 2020). ResNet uses the idea of bypassing the pathways in a deep network by adding the original input to transformed signals later in the network, as seen in Fig. 7.5. Input F_l^k is added to the transformed signal $g_c\left(F_{l\to m}^k, k_{l\to m}\right)$ and is added to the layer succeeding it, after applying the nonlinear activation. These residual blocks may involve skipping of multiple hidden layers. This results in faster convergence of the network and overcomes the diminishing gradient problem, which was one of the main problems faced in training deeper networks. This cross-layer connectivity is based on long short-term memory (LSTM)-based recurrent neural network (RNN), where two gates control the flow of information across layers. ResNet introduced the concept of residual learning in a CNN, which led to its win in the ImageNet Large Scale Visual Recognition Challenge (ILSVRC—2015) competition. Residual connections are easy to optimize and gain better accuracy even in very deep networks. When it was proposed in 2015 with a 152-layer deep network, it revolutionized the way CNNs were trained. It is 20× deeper than AlexNet and 8× compared to VGG but computationally less complex. ResNet showed 28% improved performance on image recognition benchmark dataset COCO (Lin et al. 2014).

One of the earliest open-source model proposed for the detection of COVID-19 from chest X-rays is COVID-Net that uses a tailor-made ResNet model based on user specified requirements (Wang and Wong 2020). Multiple models are generated in this study using Generative Synthesis, a machine-driven design exploration strategy, to identify the optimal micro-architecture for a given problem with specific requirements. This is achieved using a generator-inquisitor pair that would interplay to build an optimal design based on the requirement of both sensitivity and positive predictive value (PPV) for COVID-19 class ≥80%. The characteristic feature of this model is the architectural diversity, consisting of a heterogeneous mix of varying kernel sizes (7 × 7 to 1 × 1) of the convolution layers as shown in Fig. 7.6, different grouping configurations, lightweight residual projection-expansion-projection-

Table 7.1 List of popular architectures reviewed in this chapter

Literature	Mode	Model	Application	Classification	Transfer learning	Interpretability
Wang and Wong 2020	CXR	ResNet	Diagnosis	Normal/pneumonia/COVID	With ImageNet	GSInquire
Khobahi et al. 2020	CXR	FPAE, ResNet-18	Diagnosis	Normal/pneumonia/ COVID	With ImageNet	Perturbation based algorithm
Li et al. 2020	CT scans	ResNet-50, U-net	Diagnosis	COVID/CAP/non-pneumonia	No	GRAD-CAM
Gozes et al. 2020	CT scans	U-net, ResNet-50	Diagnosis/ prognosis	COVID/non-COVID	With ImageNet	GRAD-CAM
Jin et al. 2020	CT scans	DeepLab v1, ResNet-152	Diagnosis	COVID/non-COVID	With ImageNet7	Guided GRAD-CAM
Das et al. 2020	CXR	Truncated InceptionNet	Diagnosis	COVID/non-COVID/normal	With ImageNet	–
Khan et al. 2020a, b	CXR	Xception	Diagnosis	COVID/normal, Normal/pneumonia/COVID, Normal/pneumonia-bacterial/ pneumonia-viral/COVID	With ImageNet	–
Sedik et al. 2020	CXR, CT scans	CNN, ConvLSTM	Diagnosis	COVID/normal	No	--
Brunese et al. 2020	CXR	VGG-16	Diagnosis	Healthy/other pulmonary Diseases, COVID/pneumonia	With ImageNet	--
Wang et al. 2020	CT scans	DenseNet	Diagnosis/ prognosis	COVID/other pneumonia, High-risk/low-risk	With ImageNet, VESSEL12 dataset and data from patients with lung cancer	Gradient based localization method

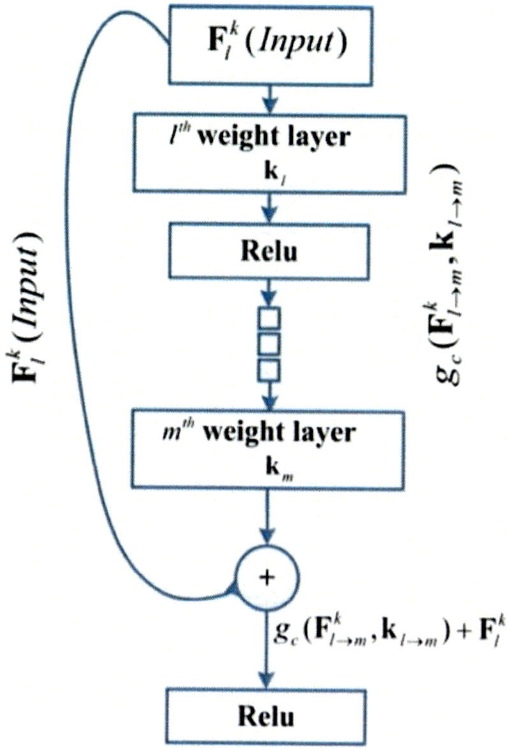

Fig. 7.5 ResNet block where the input F_l^k is added to the transformed signal $g_c\left(F_{l\to m}^k, k_{l\to m}\right)$ to enable cross-layer connectivity. (Reproduced from (Khan et al. 2020a))

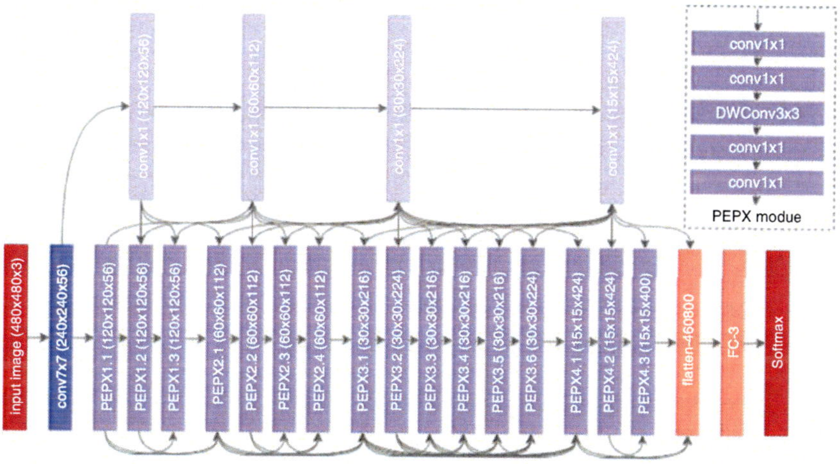

Fig. 7.6 COVID-Net architecture. (Reproduced from (Wang and Wong 2020))

extension (PEPX) design, and machine-driven selective long-range connectivity. The PEPX pattern consists of convolutions with kernel sizes 1 × 1 that project input features to lower dimensions, which expand features to higher dimensions alternately with convolutions of 3 × 3 depth, to learn spatial characteristics for minimizing computational complexity while preserving representational capacity and 1 × 1 kernels to increase the depth-wise dimensionality for obtaining final features. The long-range connections improve representational capacity, while the sparse connectivity reduces computational complexity. The COVID-Net model has been trained on a large dataset named COVIDx constructed using five different repositories and consists of ~15,000 CXR images (7966 normal, 5900 pneumonia, and 489 COVID-19 as train set and 100 images each as test set). After pre-training on ImageNet (Deng et al. 2009), the network was fine-tuned with COVIDx dataset. Pre-training a network lets the network to settle to a good starting point and will help the parameters to stabilize for general features among the pre-training data and original data, so that when they are trained on the original data, the parameters will have higher chances to optimize better. The model makes a three-class prediction—normal, non-COVID pneumonia, and COVID-19 pneumonia—and its performance is compared with two different architectures, VGG-19 and ResNet-50.

CoroNet is a DNN model that uses an autoencoder-based feature extraction network, supervised and unsupervised learning along with transfer learning (Khobahi et al. 2020). It uses a less complicated network architecture, resulting in significantly reduced number of learning parameters compared to COVID-Net (11.8 million trainable parameters, ~10 times fewer than COVID-Net), and is suitable when data is scarce. After being pre-trained on ImageNet, it is fine-tuned on a smaller subset of the dataset, COVIDx (Wang and Wong 2020). The CoroNet model comprises two modules: (1) Task Based Feature Extraction Network module (TFEN) and (2) COVID-19 Identification Network module (CIN), as shown in Fig. 7.7. TFEN is a semi-supervised module consisting of two autoencoders that generates a latent representation of the input and does an automatic segmentation of the infected areas of the lungs. The output of TFEN module and the COVID-19 data samples are fed into a classifier, CIN, for classifying the images as COVID-19 pneumonia, non-COVID pneumonia, and healthy.

COVNet is a diagnostic model that uses a supervised, 3D neural network framework for classifying 3D CT scan images as COVID-19, community-acquired pneumonia (CAP), and other lung abnormalities. Its architecture is given in Fig. 7.8. The model has the ability to extract 2D local and 3D global features. It first preprocesses the image to segment the lung region using U-Net architecture for lung segmentation and a module to identify features in a series of input CT slices, using ResNet-50 framework. A max-pooling operation then combines the features from the slices, which are then submitted to a softmax activation function through dense layers, for generating a probabilistic score for the three possible outcomes. The study was carried out on a dataset of 4352 CT scans, obtained from various hospitals in different parts of China, comprising 1292 COVID-19 samples, 1735 community-acquired pneumonia samples and other lung abnormalities CAP samples, and 1325 other lung abnormalities (Li et al. 2020).

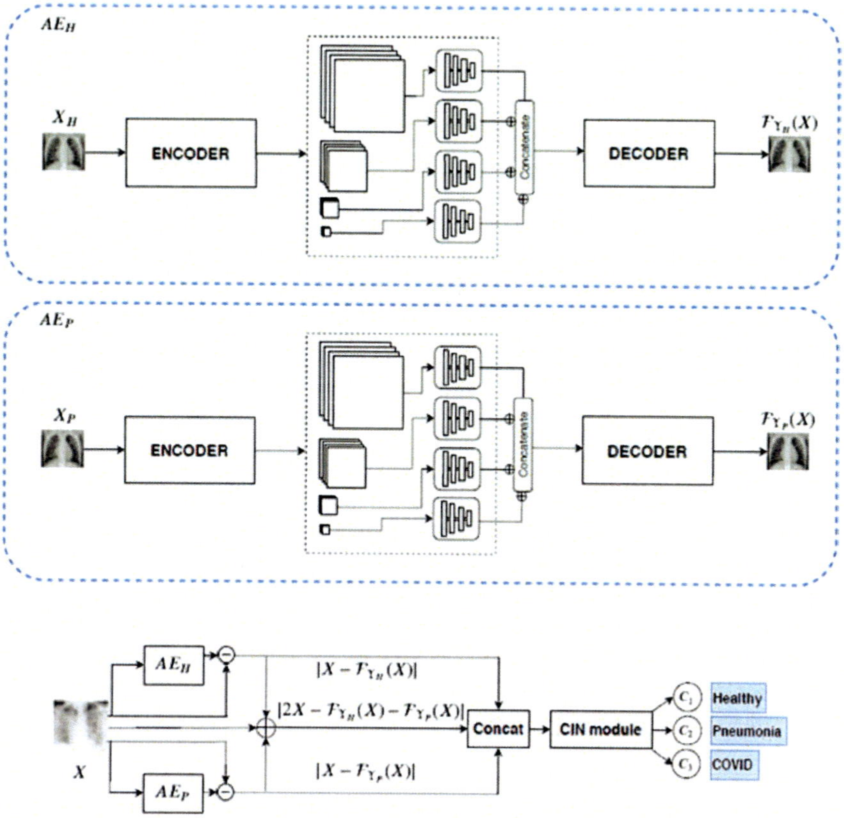

Fig. 7.7 CoroNet architecture. AE_H and AE_P are the two autoencoders trained independently on healthy and non-COVID pneumonia subjects, respectively. TFEN is a Feature Pyramid-based Autoencoder (FPAE) network, with seven layers of convolutional encoder blocks and decoder blocks, while CIN is a pre-trained ResNet-18 network. (Reproduced from (Khobahi et al. 2020))

Gozes et al. proposed an automated AI-based CT image analysis tool that utilizes 2D and 3D DL models for binary classification (COVID-19 vs non-COVID). It generates a corona score that can be utilized for measuring the progression in recovery of patients (Gozes et al. 2020). The block diagram in Fig. 7.9 for the proposed system comprises two levels: subsystems A and B. Subsystem A uses a commercial software for a 3D analysis of lung volume to detect nodules and focal opacities and provides quantitative measurement for calcification detection and texture characterization for solid vs sub-solid vs ground-glass opacities (GGO). The images that are flagged as having abnormalities from subsystem A are send through subsystem B, which performs a 2D analysis for identifying large-size diffuse opacities in each slice, clinically indicated in COVID-19 disease. A U-Net-based architecture is used to segment the lung region in this subsystem. A pre-trained ResNet-50 deep convolution neural network is used to classify images (cases per

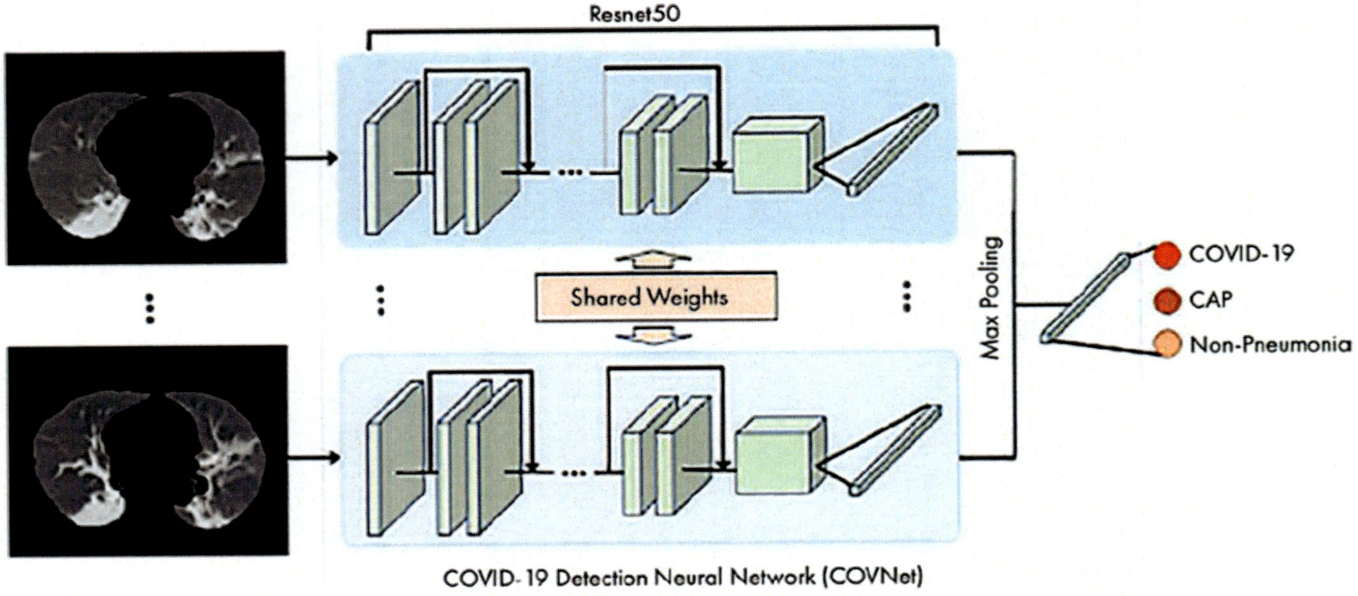

Fig. 7.8 COVNet architecture. Features are extracted from each CT scan slice which are combined using max-pooling operation and submitted to a dense layer, which generates scores for the three classes. (Reproduced from (Li et al. 2020))

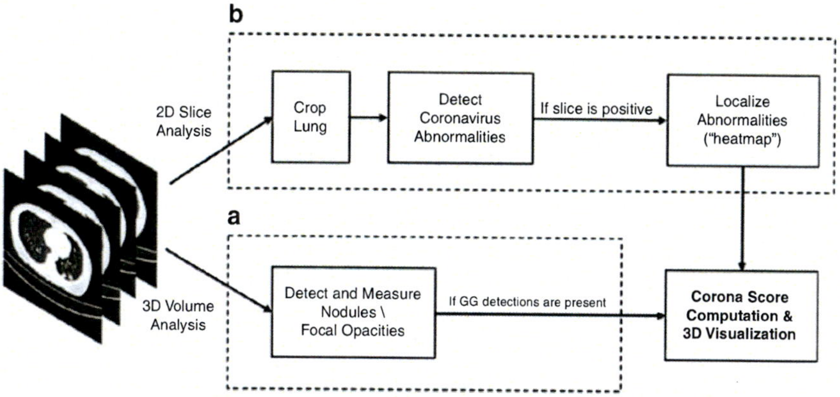

Fig. 7.9 Block diagram of the subsystem (**a**) performs a 3D analysis of CT scans, for identifying lung abnormalities, and subsystem (**b**) that performs a 2D analysis of each slice of CT scans, for detecting and marking large-sized ground-glass opacities using proposed method (reproduced from (Gozes et al. 2020))

slice) as normal vs abnormal. The model also outputs a lung abnormality localization map along with the score to identify areas contributing to the network's decision, using Grad-CAM technique (Selvaraju et al. 2017). Using this AI-based system, eight COVID-19 patients were monitored for a period of 30 days, and the relative progression of disease among the patients was assessed to allocate to the patients' required resources accordingly. Thus, AI-enabled diagnostics can help not only in the detection of but also monitoring the progression of COVID-19 disease.

Recent study by Jin et al. proposed a fast AI system based on deep neural network for diagnosis of COVID-19, by analyzing chest CT images, which achieved accuracy, sensitivity, and specificity close to 95%, and outperformed radiologists by over two orders of magnitude in diagnosis time (Jin et al. 2020). The study involved a multitask diagnostic to distinguish between COVID-19, influenza-A/B, non-viral pneumonia, and non-pneumonia cases. Here, a DL-based model trained on CT scan images, annotated by radiologists, was able to detect COVID-19 patients correctly and assist radiologists, by significantly reducing the reading time (Chen et al. 2020). The model was trained on 46,096 images from a very small set of 106 patients (51 COVID-19, 55 other conditions) retrospectively, and the prediction results were compared with the diagnosis of the radiologists. The AI system included five components: (1) a lung segmentation block, (2) COVID-19 diagnosis block, (3) a module for identifying abnormal slices in positive samples, (4) module for visualizing abnormal regions in the slices, and (5) module for explaining features of abnormal regions. Input 3D CT volumes were taken slice by slice and the lung area segmented with DeepLab v1, a 2D semantic segmentation network (Chen et al. 2016). The segmentation results were used as masks and concatenated with raw CT slices and fed to COVID diagnosis block, which has ResNet-152 as a backbone with 152 convolutional layers, pooling and dense (fully connected) layers using a 2D deep network, after pre-training it on ImageNet. Output score of this block provides confidence, whether lung-masked slices are COVID-19 positive or negative. The top three highest scores on 2D slices of a volume are averaged to obtain a 3D volume score. A block for locating the abnormal slices is the same as the diagnosis block, the difference being that it was trained only on manually curated COVID-19 positive images. The workflow of the system is depicted in Fig. 7.10a, and the dataflow in the AI system is explained in Fig. 7.10b.

7.3.1.2 Other CNN Models

Various architectures have been proposed in the analysis of X-rays and CT scans of the chest for COVID-19 and are briefly described below. ResNets, through additive identity transformations, help in solving the vanishing gradient problem. This results in many layers becoming less informative. To address this problem, models like DenseNet were introduced. Inception and VGG have proved that deeper networks are essential for solving complex tasks, like detection of ground-glass opacities (GGO) in CXR. Inception models replace large-size filters of the previous versions of CNN models with smaller-sized filters, thereby reducing the computational cost of deep networks with its performance being unaffected. The architecture of inception model is given in Fig. 7.11. Xception is another architecture that is an improvement over inception, and it introduced the idea of depth-wise separable convolution. In this model, the inception block, having different spatial dimensions (5′5, 3′3, 1′1), is replaced by a single dimension (3′3) block, followed by 1′1 convolution to reduce computational complexity (Fig. 7.12). DenseNet, similar to ResNet, uses cross-layer connectivity but with a modification of connecting feature maps of all the previous layers to all subsequent layers. Instead of adding it, DenseNet model (Fig. 7.13) concatenates the features of previous layers. Though the number of parameters in this case is very large compared to ResNet, it reduces overfitting in case of smaller datasets. VGG-19 architecture consists of 19 layers and stacking of smaller size filters (3′3) compared to filters of size (11′11, 5′5). It has a simple homogenous topology but has the drawback of training 138 million parameters (Fig. 7.14). LSTMs have the behavior of remembering information for longer periods of time, using the concept of gates controlling information flow between layers (Fig. 7.15). They are good at learning patterns from sequential data, and the new input will be weighted on their occurrence in the previous samples.

InceptionNet V3, proposed for classifying images from ImageNet, has been modified for detecting COVID-19 using CXR images (Das et al. 2020). In this proposed model, only three inception modules and one grid size module from the original architecture of Inception Net are retained along with the convolutional, pooling, and fully connected layers (Fig. 7.16). Truncation is performed to reduce complexity and avoid overfitting because of very few COVID-19 images. The Inception module consists of kernels of different receptive field sizes (1 × 1, 3 × 3, 5 × 5), compared to those of fixed field sizes in traditional CNN models; this allows it to capture features from input at multiple resolutions and of varying sizes, in parallel. A 3 × 3 max-pooled input is stacked with the output of Inception module and connected to the next convolutional layers results in the unique performance of Inception module. An adaptive learning rate is used for training with 0.001

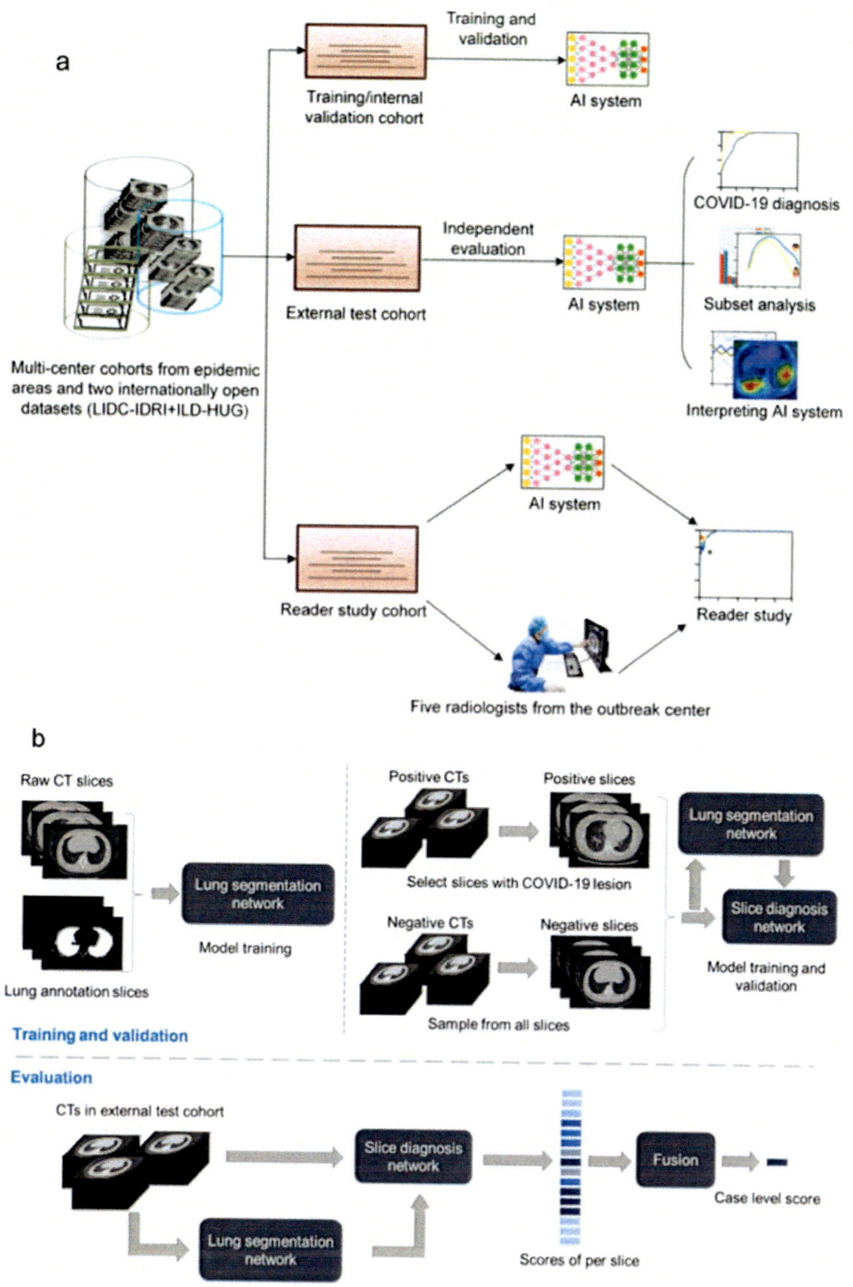

Fig. 7.10 (**a**) Workflow of the AI system data divided into four nonoverlapping cohorts for training, internal validation, external testing, and expert reader validation. (**b**) Usage of the AI system—performs lung segmentation on CT images and diagnosis of COVID-19 and locates abnormal slices (reproduced from (Jin et al. 2020))

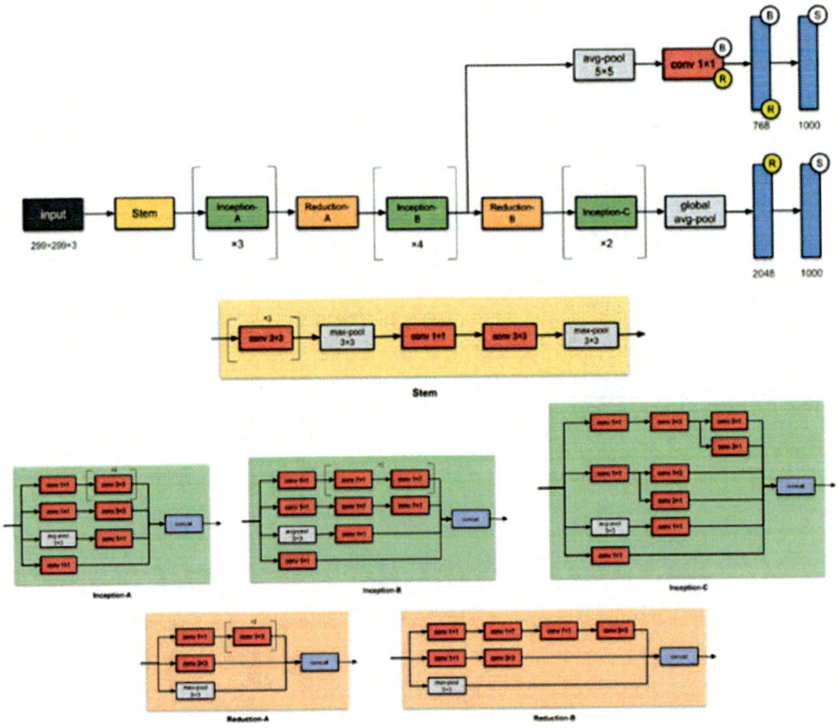

Fig. 7.11 Inception V3 architecture has a deeper architecture compared to ResNet (source https://towardsdatascience.com/illustrated-10-cnn-architectures-95d78ace614d#d27e)

as the initial value that is halved if the validation loss does not increase for three epochs, called the patience factor. To have more meaningful initial weights, the network was pre-trained on ImageNet instead of random initialization of weights. Model was trained on six different combinations of datasets of COVID-19, pneumonia, tuberculosis (TB), and healthy samples and validated using tenfold cross validation.

A deep neural network model, proposed by Khan et al., is based on Xception architecture, which is basically the extreme version of Inception. It consists of 71 layers deep CNN for classifying the CXR images into binary (COVID-19 and normal), three class (COVID-19, normal and pneumonia), and four class (COVID, normal, pneumonia-bacterial, and pneumonia-viral) classifications (Khan et al. 2020b). It uses depth-wise separable convolution layers along with residual connections, replacing $n \times n \times n$ convolutions with $1 \times 1 \times k$ point-wise convolutions, followed by channel-wise $n \times n$ spatial convolution operations, resulting in reducing the number of operations by a factor $1/k$. In this case, also the network was pre-trained on ImageNet and fine-tuned on the task-specific dataset for 80 epochs in batches of 10. Softmax activation function was applied on the output of the last connected layer and probability distribution generated for the

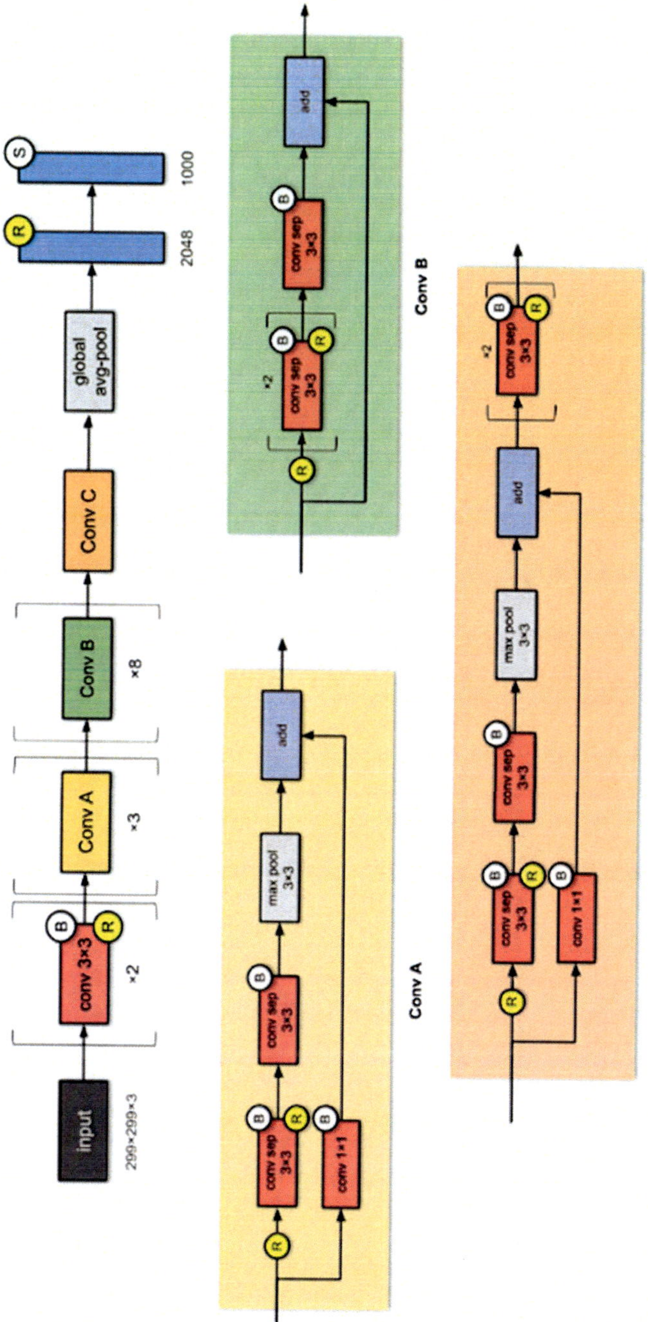

Fig. 7.12 Xception architecture introduced depth-wise separable convolutions (source https://towardsdatascience.com/illustrated-10-cnn-architectures-95d78ace614d#d27e)

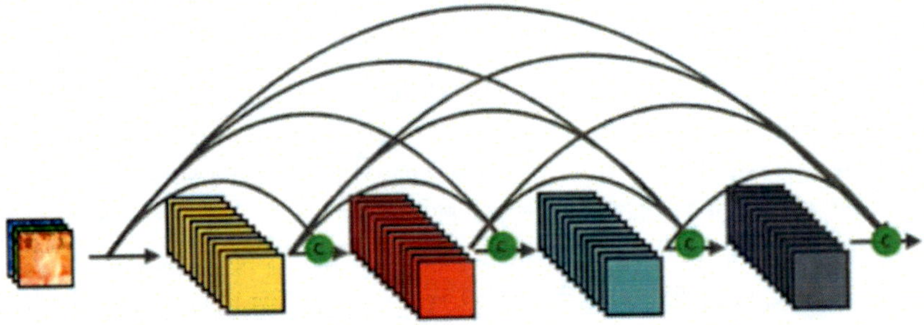

Fig. 7.13 DenseNet architecture connects feature maps of all previous layers to subsequent layers (source https://towardsdatascience.com/review-densenet-image-classification-b6631a8ef803)

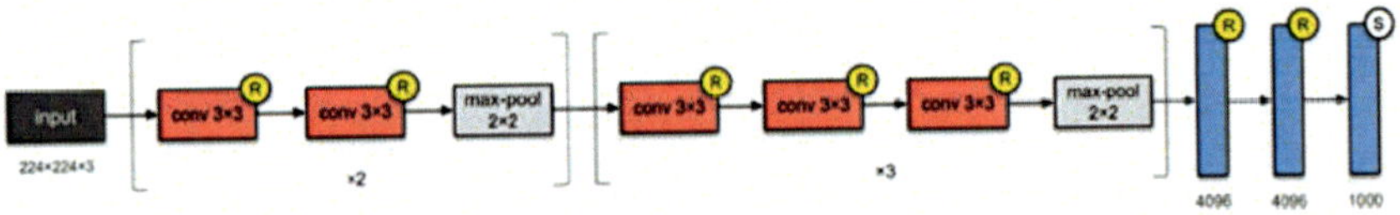

Fig. 7.14 VGG architecture has a narrow topology (source https://towardsdatascience.com/illustrated-10-cnn-architectures-95d78ace614d#d27e)

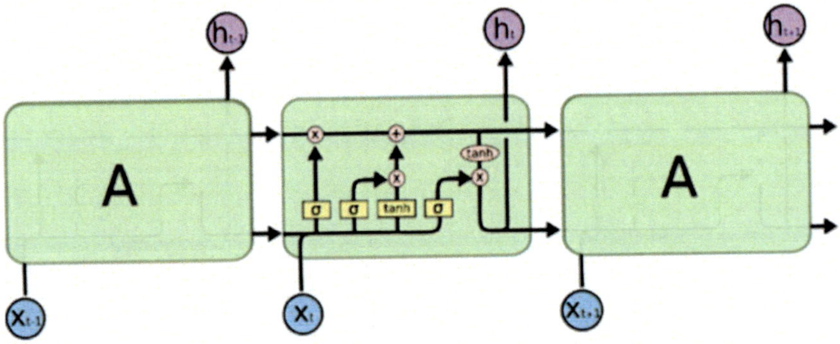

Fig. 7.15 LSTM architecture employs gates to regulate flow of information across layers (source http://colah.github.io/posts/2015-08-Understanding-LSTMs/)

output classes. Performance of the network was estimated using fourfold cross validation. To address data imbalance, random under-sampling, was done by randomly deleting samples in majority classes.

The diagnostic analysis on both CXR and CT images using two data augmented DL models, CNN and convoluted long short-term memory (ConvLSTM), has been conducted by Sedik et al. (2020). LSTMs have an architecture similar to ResNets with respect to the cross-layer connectivity. The difference is that in LSTM, it is done with the help of gates that control the information that is passed over the layers. For the data augmentation process, a set of simple image transformations, such as

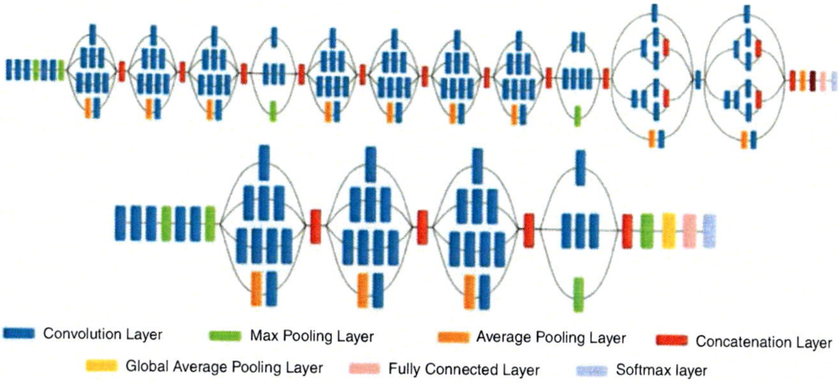

Fig. 7.16 @Original Inception Net Architecture (above), truncated Inception Net architecture (below). (Reproduced from (Das et al. 2020))

scaling, rotation, and flipping resulted in a tenfold increase and was followed by convolutional generative adversarial networks (CGANs) to address the problem of limited data. During the learning phase on a given training set, GANs generate new synthetic data from the existing ones for training. It is composed of generator and discriminator networks. The generator synthesizes new data from the latent space of the input, while the discriminator tries to distinguish reconstructed images from the input images. The generator's objective is to increase discriminator's error rate. In this study, CGAN consisted of five convolutional transpose layers with filters of sizes 8, 4, 2, 1, and 1, respectively, in the generator (Fig. 7.17). Input is given to a denoising fully connected layer, followed by convolutional transpose layers and batch normalization layers. At the end of generator, feature maps of input images are generated. The discriminator consisted of five convolutional layers with filters of sizes 64, 2, 4, 8, and 1, respectively, followed by Conv2D layers, batch normalization layers, and a denoising fully connected layer. All the images were resized before feature extraction. The two deep learning models, CNN and ConvLSTM, comprising five and one convolution layers, respectively, are followed by max-pooling and global average pooling (GAP) layers for determining and extracting the features, which are then fed to the classifier. The performance of classifiers is carried out with and without data augmentation, to assess its role in diagnosing COVID-19 using support vector machine and k-nearest neighbor classifiers to compare with traditional ML techniques. As expected, the DL models exhibited better performance than SVM and k-NN models.

The study by Brunese et al. proposed a DL model based on VGG-16 (i.e., Visual Geometry Group) architecture, a CNN with 16 layers (Brunese et al. 2020). The model works in three phases: Initially, it tries to distinguish between healthy and pneumonia-related images. In the next phase, it attempts to differentiate between COVID-19 images and other pneumonia, and finally, the last phase identifies regions in the image that are symptomatic of COVID-19 to provide an explainable system. The architecture used in this study is shown in Fig. 7.18. To exploit transfer learning,

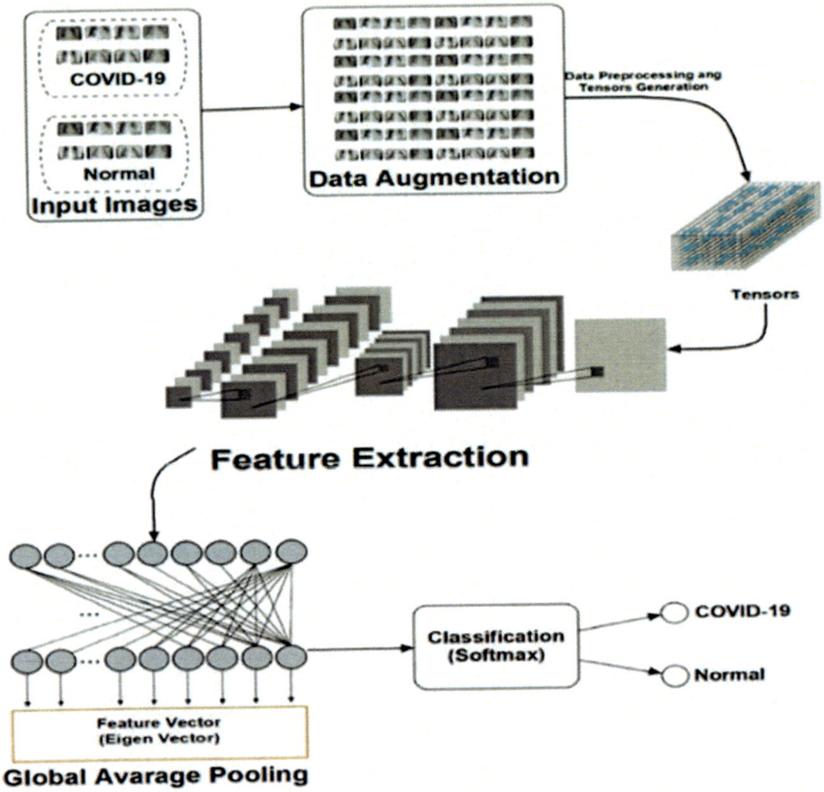

Fig. 7.17 Dataflow in the DL model using data augmentation (reproduced from (Sedik et al. 2020))

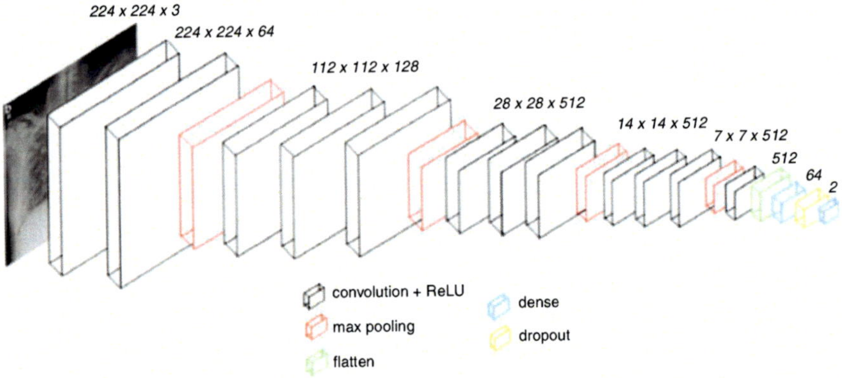

Fig. 7.18 Architecture used in the study by (reproduced from (Brunese et al. 2020))

the model has been pre-trained with over 14 million images from ImageNet. To fine-tune transfer learning, the network is appended with a new fully connected (FC) layer, after removing the FC layer from pre-trained network, and image resized to 224 × 224 dimensions, the size of the first convolutional layer. This new FC layer includes AveragePooling2D (which performs the average pooling operation by computing the average of every patch in the feature map), Flatten (the input is flattened by transforming a 2D feature matrix into a vector that is given as an input to a classifier), Dense (to transform data, in this case to reduce the height of the vector from 512 to 64 elements), Dropout (to build a network that does not overfit to training data by randomly selecting neurons not used during training), and another Dense connection (this dense reduces the height of the vector from 64 to 2, corresponding to the two classes to predict) with softmax function used for classification (Fig. 7.18). For data generalization, data augmentation was carried out with a random clockwise or counterclockwise rotation of 15 degrees. For the explainability feature, gradient-weighted class activation mapping (Grad-CAM) approach is used here, which results in a heatmap depicting abnormal regions for each class. The model training, validation, and testing were done on a dataset with 6523 CXR images, which includes 3520 related to normal class, 2753 related to other pulmonary-related illness, and 250 COVID-19 images, confirmed by radiologists.

Wang et al. proposed a DL model, COVID-19Net, for diagnostic and prognostic analysis of CT images (Wang et al. 2020). The unique feature of this study is using retrospective CT images for pre-training the network, to learn the lung features, and using follow-up data of 5+ days of 471 COVID-19 patients for training and validating the model. That is, the transfer learning here is done on chest CT images, compared to ImageNet database used in most other studies. The model can identify the patients with high-risk from low-risk group based on follow-up data, allowing for early interventions and resource management. Through a prognostic feature selection procedure, three features, namely, age, sex, and comorbidity, along with a 64 DL model-generated features were combined to predict a hazard value for each patient using a multivariate Cox proportional hazard (CPH) model. The median of this cutoff value was used for dividing patients into low-risk and high-risk groups. The model included three modules for (1) lung segmentation, (2) non-lung area suppression, and (3) COVID-19 diagnosis and prognosis. The architecture and dataflow in the model are depicted in Fig. 7.19. Lung segmentation module was build using DenseNet121-FPN, pre-trained on ImageNet and fine-tuned on VESSEL12 dataset. Through this procedure, lung mask in the CT image is obtained, and the lung ROI was extracted using a bounding box from the CT images. This may include non-lung areas, e.g., the heart, spine, etc., which were removed by suppressing the intensities of these areas. The final lung ROI was standardized using z-score normalization before being fed to COVID-19Net. The model uses DenseNet-like structure with four dense blocks, each having multiple stacked convolutional, batch normalization, and ReLU activation layers. The dense connections in each layer were used to capture multilevel information from the images. A global average pooling layer added at the end of convolutional layer

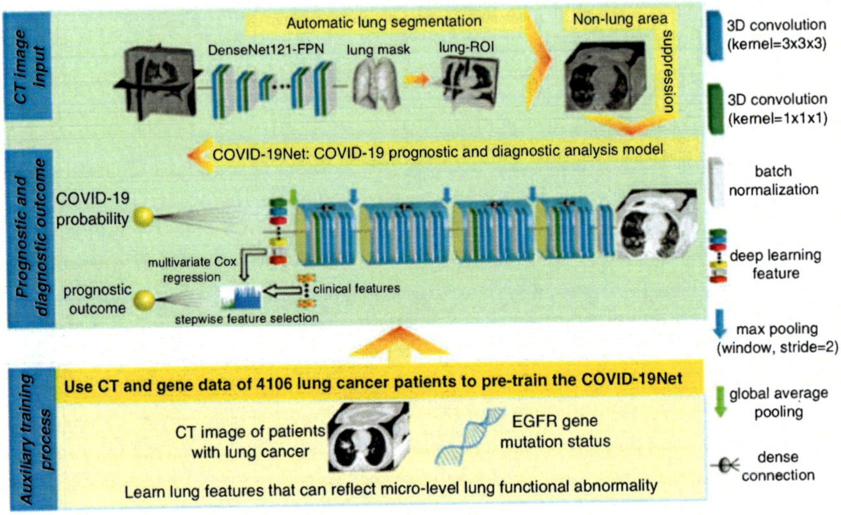

Fig. 7.19 Illustration of the COVID-19Net model (reproduced from (Wang et al. 2020))

generates a 64-dimentional feature vector, which is connected to the output neuron to predict the probability of a patient having COVID-19 infection.

7.3.2 The Data Imbalance Challenge

Machine learning models rely on data to learn the patterns in the data. The amount of data needed depends on how deeply connected are the features in the data. The nonlinear relationships in the underlying patterns can only be captured by a model, if it has been trained sufficiently with data possessing all these features. In the case of DL models, this is a stringent requirement. The major bottleneck in the analysis of COVID-19 image data has been limited availability of public data compared to other lung infections, leading to class imbalance problem. To overcome these challenges, most studies presented here used transfer learning approach. Transfer learning (TL) is a technique in which model trained on one data is used for initializing the parameters of another related problem. Given a source domain and target domain with respective source task and destination task, learning conditional probability distribution in target domain with insight gained from source domain is the objective of this approach. This technique has been applied in various classification tasks, such as classification of cancer samples (Sevakula et al. 2019), detection/classification of Alzheimer's disease (Hon and Khan 2017; Maqsood et al. 2019), etc. The major advantage of using transfer learning is reduced time for training a neural network model and may also result in better generalization of the model.

Image analysis models are typically trained on ImageNet dataset, which consists of over 14 million images organized under more than 100,000 "synonym sets" or "synsets." The advantage of training a neural network using ImageNet is to have a better starting point for learning a new task compared to random initializations. Many of the studies discussed above, namely, COVID-Net (Wang and Wong 2020), CoroNet (Khobahi et al. 2020), Gozes et al., AI system by Jin et al. (2020), and CoroNet model by Khan et al., have all been trained on ImageNet database. In all these cases, the knowledge gained by pre-training on ImageNet is used to identify images with pulmonary-related diseases. In COVID-19Net, the pre-training was done on chest CT scans of lung cancer patient data, for which epidermal growth factor receptor (EGFR) gene sequencing data were available. This enabled the DL model to learn lung features associated with lung abnormalities.

7.3.3 Interpretation/Visualization of Results

From the recent literature, it is evident that though deep learning models have attained unparalleled accuracy in the classification and segmentation of images, their major limitation is interpretability, that is, to identify the features responsible in decision-making. This is the most important component in model understanding and model debugging and has limited the acceptance of DL methods, especially in the medical field. Radiologists, using AI models to assist them in diagnosing various conditions in a health support system, need to reaffirm the decisions that they make using these models. This requires the model to help them interpret or visualize the features that enabled the model's decision. In COVID-19 diagnosis using chest CXR, it is very crucial that model is rightly distinguishing symptomatic COVID-19 pneumonia from other lung infections. This requires marking on the lung images the regions that helped in differentiating between different pneumonia-related cases. Different methods have been proposed in the studies discussed above for identifying and visualizing the features responsible for prediction. For example, COVID-Net makes use of GSInquire method, which projects the updated parameters by the inquisitor to improve the network generated by the generator into the same subspace as the input x. This helps in visualizing the pixel areas that contributed for the prediction of the label as shown in red in Fig. 7.20 and helps in confirming that the algorithm is not making incorrect decisions based on imaging artifacts, etc. Thus, apart from providing insights into the factors associated with COVID-19, this would help the clinicians in the screening process with improved accuracy.

The deep learning model CoroNet (Khobahi et al. 2020) uses attribution map, which is basically a heatmap, showing the pixels that contributed to the prediction. These attribution maps are generated using a perturbation-based algorithm and shown in Fig. 7.21 for three categories: normal, pneumonia, and COVID-19. The algorithm works by perturbing the input image, such that the target class probability is minimized. The pixels that minimize the target class probability by a great extent are highlighted on the heatmaps as the regions contributing to the correct predictions. It maybe be noted that different regions in the CXR images are highlighted by the classification model for the three categories considered.

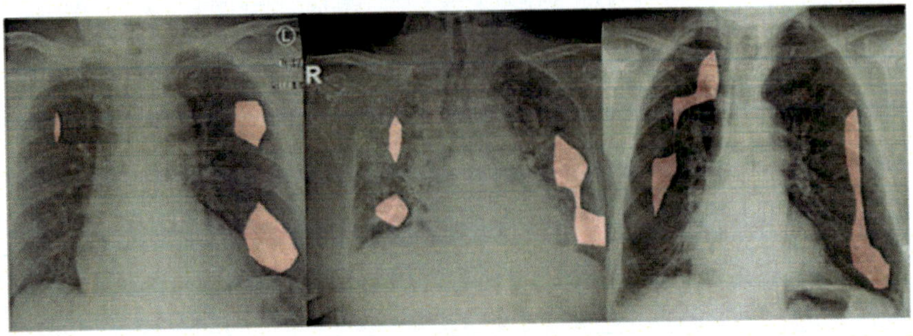

Fig. 7.20 Abnormal lung regions identified by GSInquire leveraged from the update parameters generated by the Inquisitor of the generator-inquisitor pair after probing the response signals from the generated network with respect to the input signal and target label. (Reproduced from (Wang and Wong 2020))

The DL model COVNet (Li et al. 2020) uses a Grad-CAM to generate heatmaps for visualizing the areas associated with the three prediction categories: COVID-19, community-acquired pneumonia (CAP), and non-pneumonia. The heatmaps are then overlapped with the original images as seen in Fig. 7.22, and red regions are associated with the predicted classes. The visualization method used in COVID-19Net (Wang et al. 2020) was based on the gradient-based localization method. The proposed DL system identified the inflammatory areas as suspicious lung areas as shown in Fig. 7.23. The regions exhibiting lesions with consolidation, ground-glass opacity (GGO), diffuse, or mixture patterns were automatically identified in agreement with the radiologist's observations in COVID-19 patients. The DL model developed by Gozes et al. (2020) also uses Grad-CAM technique to generate network activation maps. Overlap of activation maps with diffused opacities clearly shows the network's learning abilities and providing visual explanations to the predictions made. The quantitative opacity measurements and the visualization of larger opacities were based on slice-level heatmaps. To explain the results of their AI system, Jin et al. (2020) used a guided Grad-CAM for visualizing the abnormal regions in the CT images associated with COVID-19 diagnosis. These features were found to be consistent with the anatomical findings of COVID-19. The predictions were also confirmed with the readings of five expert radiologists and were mostly in agreement. GRAD-CAM algorithm was also used by Sedik et al. (2020) in localizing the areas that the DL network used for its prediction. The algorithm uses the gradients of output with respect to the final convolutional layer and outputs a coarse localization map that highlights the areas used by the network for prediction, which were consistent with the areas marked by radiologists.

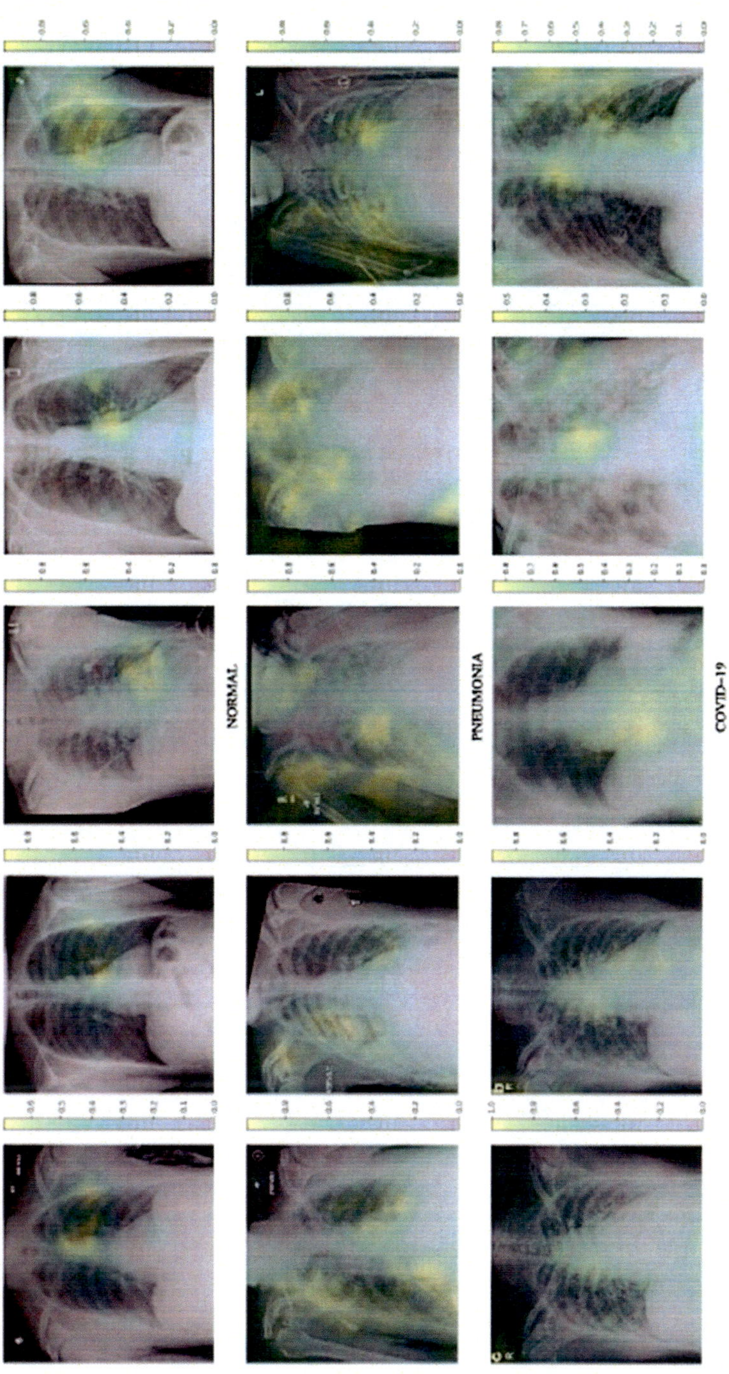

Fig. 7.21 Attribution maps for five random patients for the three classifications considered. Yellow regions represent most salient and blue regions the least salient regions as indicated by the color bar (reproduced from (Khobahi et al. 2020))

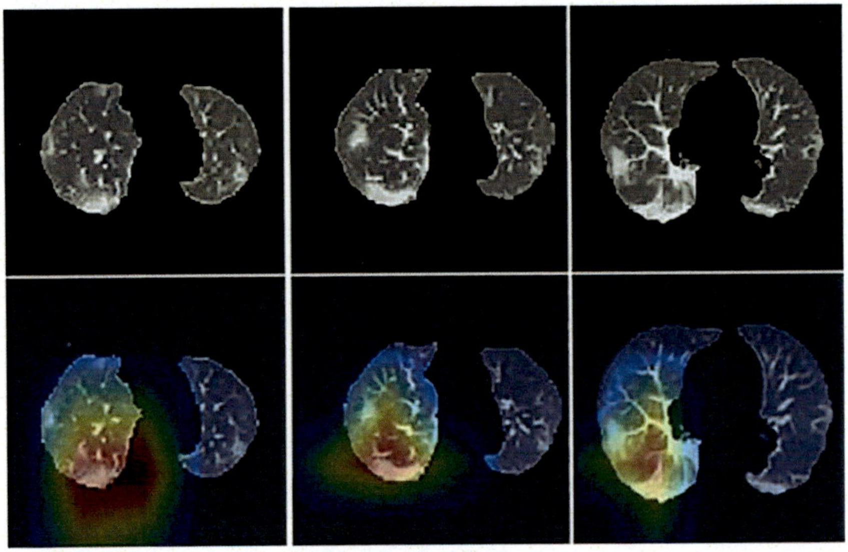

Fig. 7.22 Attention heatmaps generated by GRAD-CAM. The red regions indicate the activation regions associated with a sample. (Reproduced from (Li et al. 2020))

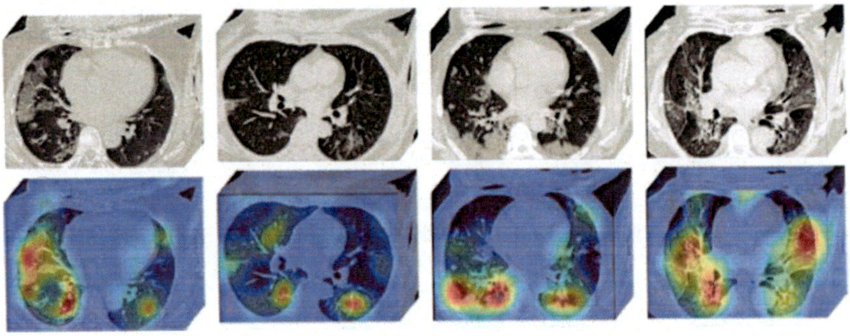

Fig. 7.23 DL discovered suspicious lung areas learned by COVID-19Net. (Reproduced from (Wang et al. 2020))

7.3.4 Performance Measurement Metrics

In most ML-based studies, accuracy, sensitivity, specificity, precision, F1-score, area under the ROC curve, etc. are the most commonly used in evaluating the model's performance and are defined below.

$$\text{accuracy} = \frac{T_p + T_n}{T_p + T_n + F_p + F_n} \tag{7.1}$$

$$\text{sensitivity} = \frac{T_p}{T_p + F_n} \tag{7.2}$$

$$\text{specificity} = \frac{T_n}{T_n + F_p} \tag{7.3}$$

$$\text{precision} = \frac{T_p}{T_p + F_p} \tag{7.4}$$

where T_p is defined as true positive, T_n as true negative, F_p as false positive, and F_n as false negative.

$$F1\ \text{score} = 2 \times \frac{(\text{precision} \times \text{sensitivity})}{(\text{precision} + \text{sensitivity})} \tag{7.5}$$

Traditionally, the models' performance is most commonly evaluated using accuracy. However, for applications, where high imbalance of classes in data is observed, accuracy may not be a suitable metric, because even when the model predicts the entire test samples into a single class, accuracy would still be high, giving a false impression of the model's performance. In such situations, other metrics, e.g., sensitivity, specificity, F1-score, area under the ROC curve, etc., can be considered, which give a better picture of the model's performance. Sensitivity is a very crucial measure in case of medical applications, because a good sensitivity score indicates that the model does not miss any positive samples. Equally important is precision, because a good precision score indicates that the model does not misclassify a negative sample and cause mental trauma to patients and waste hospital resources in such pandemic situations. The performance measurement is usually done using k-cross validation technique, wherein the dataset is divided into k sets and $k - 1$ that are used for training, and k^{th} set is used for testing. This is repeated recursively until all k sets have been used for testing. This technique helps to avoid any bias in the training or testing samples and can handle the problems associated with outliers. Almost all the studies discussed in this chapter has shown good performances of their proposed models in terms of accuracy, sensitivity, specificity, recall, etc.

7.4 Challenges

CNNs have achieved great performances in many challenging tasks, but there are still grey areas in its performance when it comes to its application in certain areas, such as medical domain. Data is the backbone of any ML tool, especially for supervised learning algorithms, and the lack of annotated data is the most challenging, among other reasons that ML researchers face toward making the tool confident in assisting the healthcare professionals. The largest available number of patients CXR or CT images are still very small despite the increase in number of cases worldwide. For training a ML tool with complex patterns like ground-glass opacities in case of chest images, the minimum requirement is balanced training data, which is a far cry from the reality. Deep neural networks are typically considered a black box when it comes to the explainability of its results. Visualization of results from DL models, interpreting the predictions with good precision and confidence, is the need of the hour and needs to be addressed. Further, these technologies have to be made available in portable devices like smartphones, so that the objectives of the research are materialized. The training of deep networks requires powerful computational resources, which makes it challenging to embed them in smaller portable devices.

7.5 Summary

COVID research is moving at faster than before rates, and thousands of new research publications have come in the past few months. This chapter has focused on reviewing few DL-based solutions for diagnosing COVID-19, using chest radiology images. Most of the studies have exploited the capability of CNNs to bring out a reliable diagnostic/prognostic tool analyzing CXR or CT scans of the chest. ResNet, DenseNet, Inception, Xception, VGG, and other customized models have emerged as forerunners in this task. The performances of all these models are comparable, and most of them give very high accuracy, sensitivity, and specificity. The visualization of the results of these models is also presented as part of the performance of these models. Most studies have used attention heatmaps to visualize the activation regions in the images that resulted in model prediction. GRAD-CAM is one such technology in generating attention heatmaps. Localization of abnormal lung regions are also addressed that highlight only the lung regions responsible for model's decision from the entire image, proving the correctness of the methods. Though the results from these studies are promising, generalizability of these models on data from different distributions need to be verified.

References

Ai T, Yang Z, Hou H, Zhan C, Chen C, Lv W, Tao Q, Sun Z, Xia L (2020) Correlation of chest CT and RT-PCR testing in coronavirus disease 2019 (COVID-19) in China: a report of 1014 cases. Radiology. https://doi.org/10.1148/radiol.2020200642

Antin B, Kravitz J, Martayan E (2017) Detecting pneumonia in chest X-rays with supervised learning. Semanticscholar.org

Awulachew E, Diriba K, Anja A, Getu E, Belayneh F (2020) Computed tomography (CT) imaging features of patients with COVID-19: systematic review and meta-analysis. Radiol Res Pract. https://doi.org/10.1155/2020/1023506

Bressem KK, Adams LC, Erxleben C, Hamm B, Niehues SM, Vahldiek JL (2020) Comparing different deep learning architectures for classification of chest radiographs. Sci Rep 10:13590. https://doi.org/10.1038/s41598-020-70479-z

Brunese L, Mercaldo F, Reginelli A, Santone A (2020) Explainable deep learning for pulmonary disease and coronavirus COVID-19 detection from X-rays. Comput Methods Prog Biomed 196:105608. https://doi.org/10.1016/j.cmpb.2020.105608

Chandra TB, Verma K (2020) Pneumonia detection on chest X-ray using machine learning paradigm. In: Chaudhuri BB, Nakagawa M, Khanna P, Kumar S (eds) Proceedings of 3rd international conference on computer vision and image processing, advances in intelligent systems and computing. Springer, Singapore, pp 21–33. https://doi.org/10.1007/978-981-32-9088-4_3

Chen L-C, Papandreou G, Kokkinos I, Murphy K, Yuille AL (2016) Semantic image segmentation with deep convolutional nets and fully connected CRFs. arXiv:1412.7062 [cs]

Chen J, Wu L, Zhang J, Zhang L, Gong D, Zhao Y, Hu S, Wang Y, Hu X, Zheng B, Zhang K, Wu H, Dong Z, Xu Y, Zhu Y, Chen X, Yu L, Yu H (2020) Deep learning-based model for detecting 2019 novel coronavirus pneumonia on high-resolution computed tomography: a prospective study. medRxiv 2020.02.25.20021568. https://doi.org/10.1101/2020.02.25.20021568

Cong L, Ran FA, Cox D, Lin S, Barretto R, Habib N, Hsu PD, Wu X, Jiang W, Marraffini LA, Zhang F (2013) Multiplex genome engineering using CRISPR/Cas systems. Science 339:819–823. https://doi.org/10.1126/science.1231143

Das D, Santosh KC, Pal U (2020) Truncated inception net: COVID-19 outbreak screening using chest X-rays. Phys Eng Sci Med 43(3):915–925. https://doi.org/10.1007/s13246-020-00888-x

Deng J, Dong W, Socher R, Li L-J, Li K, Fei-Fei L (2009) ImageNet: a large-scale hierarchical image database. In: 2009 IEEE conference on computer vision and pattern recognition. Presented at the 2009 IEEE conference on computer vision and pattern recognition. IEEE, pp 248–255. https://doi.org/10.1109/CVPR.2009.5206848

Fang Y, Zhang H, Xie J, Lin M, Ying L, Pang P, Ji W (2020) Sensitivity of chest CT for COVID-19: comparison to RT-PCR. Radiology 296:E115–E117. https://doi.org/10.1148/radiol.2020200432

Gozes O, Frid-Adar M, Greenspan H, Browning PD, Zhang H, Ji W, Bernheim A, Siegel E (2020) Rapid AI development cycle for the coronavirus (COVID-19) pandemic: initial results for automated detection & patient monitoring using deep learning CT image analysis. arXiv:2003.05037 [cs, eess]

Guo J, He H, He T, Lausen L, Li M, Lin H, Shi X, Wang C, Xie J, Zha S, Zhang A, Zhang H, Zhang Z, Zhang Z, Zheng S, Zhu Y(2020) GluonCV and GluonNLP: deep learning in computer vision and natural language processing. arXiv:1907.04433 [cs, stat]

He K, Zhang X, Ren S, Sun J (2015) Deep residual learning for image recognition. arXiv:1512.03385 [cs]

Hon M, Khan N (2017) Towards Alzheimer's disease classification through transfer learning. arXiv:1711.11117 [cs]

Hua K-L, Hsu C-H, Hidayati SC, Cheng W-H, Chen Y-J (2015) Computer-aided classification of lung nodules on computed tomography images via deep learning technique. Onco Targets Ther 8:2015–2022. https://doi.org/10.2147/OTT.S80733

Huang G, Liu Z, van der Maaten L, Weinberger KQ (2018) Densely connected convolutional networks. arXiv:1608.06993 [cs]

Islam MT, Aowal MA, Minhaz AT, Ashraf K (2017) Abnormality detection and localization in chest x-rays using deep convolutional neural networks. arXiv:1705.09850 [cs]

Jin C, Chen W, Cao Y, Xu Z, Tan Z, Zhang X, Deng L, Zheng C, Zhou J, Shi H, Feng J (2020) Development and evaluation of an AI system for COVID-19 diagnosis. medRxiv 2020.03.20.20039834. https://doi.org/10.1101/2020.03.20.20039834

Khan A, Sohail A, Zahoora U, Qureshi AS (2020a) A survey of the recent architectures of deep convolutional neural networks. Artif Intell Rev. https://doi.org/10.1007/s10462-020-09825-6

Khan AI, Shah JL, Bhat MM (2020b) CoroNet: a deep neural network for detection and diagnosis of COVID-19 from chest x-ray images. Comput Methods Prog Biomed 196:105581. https://doi.org/10.1016/j.cmpb.2020.105581

Khobahi S, Agarwal C, Soltanalian M (2020) CoroNet: a deep network architecture for semi-supervised task-based identification of COVID-19 from chest x-ray images. medRxiv 2020.04.14.20065722. https://doi.org/10.1101/2020.04.14.20065722

Kucirka LM, Lauer SA, Laeyendecker O, Boon D, Lessler J (2020) Variation in false-negative rate of reverse transcriptase polymerase chain reaction–based SARS-CoV-2 tests by time since exposure. Ann Intern Med. https://doi.org/10.7326/M20-1495

Lakhani P, Sundaram B (2017) Deep learning at chest radiography: automated classification of pulmonary tuberculosis by using convolutional neural networks. Radiology 284:574–582. https://doi.org/10.1148/radiol.2017162326

Li L, Qin L, Xu Z, Yin Y, Wang X, Kong B, Bai J, Lu Y, Fang Z, Song Q, Cao K, Liu D, Wang G, Xu Q, Fang X, Zhang S, Xia J, Xia J (2020) Using artificial intelligence to detect COVID-19 and community-acquired pneumonia based on pulmonary CT: evaluation of the diagnostic accuracy. Radiology 296:E65–E71. https://doi.org/10.1148/radiol.2020200905

Lin T-Y, Maire M, Belongie S, Hays J, Perona P, Ramanan D, Dollár P, Zitnick CL (2014) Microsoft COCO: common objects in context. In: Fleet D, Pajdla T, Schiele B, Tuytelaars T (eds) Computer vision – ECCV 2014, lecture notes in computer science. Springer, Cham, pp 740–755. https://doi.org/10.1007/978-3-319-10602-1_48

Maqsood M, Nazir F, Khan U, Aadil F, Jamal H, Mehmood I, Song O (2019) Transfer learning assisted classification and detection of Alzheimer's disease stages using 3D MRI scans. Sensors (Basel) 19. https://doi.org/10.3390/s19112645

Mehendale N (2020) Facial emotion recognition using convolutional neural networks (FERC). SN Appl Sci 2:446. https://doi.org/10.1007/s42452-020-2234-1

Mojica FJM, Díez-Villaseñor C, García-Martínez J, Almendros C (2009) Short motif sequences determine the targets of the prokaryotic CRISPR defence system. Microbiology (Reading) 155:733–740. https://doi.org/10.1099/mic.0.023960-0

Nagpal S, Singh M, Singh R, Vatsa M (2019) Deep learning for face recognition: pride or prejudiced? arXiv:1904.01219 [cs]

Ng M-Y, Lee EY, Yang J, Yang F, Li X, Wang H, Lui MM, Lo CS-Y, Leung B, Khong P-L, Hui CK-M, Yuen K, Kuo MD (2020) Imaging profile of the COVID-19 infection: radiologic findings and literature review. Radiol Cardiothorac Imaging 2:e200034. https://doi.org/10.1148/ryct.2020200034

Pérez-Hernández F, Tabik S, Lamas A, Olmos R, Fujita H, Herrera F (2020) Object detection binary classifiers methodology based on deep learning to identify small objects handled similarly: application in video surveillance. Knowl-Based Syst 194:105590. https://doi.org/10.1016/j.knosys.2020.105590

Rajpurkar P, Irvin J, Zhu K, Yang B, Mehta H, Duan T, Ding D, Bagul A, Langlotz C, Shpanskaya K, Lungren MP, Ng AY (2017) CheXNet: radiologist-level pneumonia detection on chest x-rays with deep learning. arXiv:1711.05225 [cs, stat]

Ren S, He K, Girshick R, Sun J (2016) Faster R-CNN: towards real-time object detection with region proposal networks. arXiv:1506.01497 [cs]

Sedik A, Iliyasu AM, Abd El-Rahiem B, Abdel Samea ME, Abdel-Raheem A, Hammad M, Peng J, Abd El-Samie FE, Abd El-Latif AA (2020) Deploying machine and deep learning models for efficient data-augmented detection of COVID-19 infections. Viruses 12:769. https://doi.org/10.3390/v12070769

Selvaraju RR, Cogswell M, Das A, Vedantam R, Parikh D, Batra D (2017) Grad-CAM: visual explanations from deep networks via gradient-based localization. In: 2017 IEEE international conference on computer vision (ICCV). Presented at the 2017 IEEE international conference on computer vision (ICCV). IEEE, pp 618–626. https://doi.org/10.1109/ICCV.2017.74

Sevakula RK, Singh V, Verma NK, Kumar C, Cui Y (2019) Transfer learning for molecular cancer classification using deep neural networks. IEEE/ACM Trans Comput Biol Bioinform 16 (6):2089–2100. https://doi.org/10.1109/TCBB.2018.2822803

Smailagic A, Costa P, Gaudio A, Khandelwal K, Mirshekari M, Fagert J, Walawalkar D, Xu S, Galdran A, Zhang P, Campilho A, Noh HY (2020) O-MedAL: online active deep learning for medical image analysis. WIREs Data Min Knowl Discov 10:e1353. https://doi.org/10.1002/widm.1353

Spanhol FA, Oliveira LS, Petitjean C, Heutte L (2016) Breast cancer histopathological image classification using convolutional neural networks. In: 2016 international joint conference on neural networks (IJCNN). Presented at the 2016 international joint conference on neural networks (IJCNN). IEEE, pp 2560–2567. https://doi.org/10.1109/IJCNN.2016.7727519

Szegedy C, Vanhoucke V, Ioffe S, Shlens J, Wojna Z (2015) Rethinking the inception architecture for computer vision. arXiv:1512.00567 [cs]

Thi VLD, Herbst K, Boerner K, Meurer M, Kremer LP, Kirrmaier D, Freistaedter A, Papagiannidis D, Galmozzi C, Stanifer ML, Boulant S, Klein S, Chlanda P, Khalid D, Miranda IB, Schnitzler P, Kräusslich H-G, Knop M, Anders S (2020) A colorimetric RT-LAMP assay and LAMP-sequencing for detecting SARS-CoV-2 RNA in clinical samples. Sci Transl Med 12. https://doi.org/10.1126/scitranslmed.abc7075

Varshni D, Thakral K, Agarwal L, Nijhawan R, Mittal A (2019) Pneumonia detection using CNN based feature extraction. In: 2019 IEEE international conference on electrical, computer and communication technologies (ICECCT). Presented at the 2019 IEEE international conference on electrical, computer and communication technologies (ICECCT). IEEE, pp 1–7. https://doi.org/10.1109/ICECCT.2019.8869364

Waleed Salehi A, Baglat P, Gupta G (2020) Review on machine and deep learning models for the detection and prediction of coronavirus. Mater Today Proc. https://doi.org/10.1016/j.matpr.2020.06.245

Wang L, Wong A (2020) COVID-Net: a tailored deep convolutional neural network design for detection of COVID-19 cases from chest x-ray images. arXiv:2003.09871 [cs, eess]

Wang S, Zha Y, Li W, Wu Q, Li X, Niu M, Wang M, Qiu X, Li H, Yu H, Gong W, Bai Y, Li L, Zhu Y, Wang L, Tian J (2020) A fully automatic deep learning system for COVID-19 diagnostic and prognostic analysis. Eur Respir J. https://doi.org/10.1183/13993003.00775-2020

Applications of Machine Learning Algorithms in Cancer Diagnosis

8

Abstract

Cancer is a deadly disease and is a leading cause of death worldwide. It is a heterogeneous disease with a number of subtypes. For an effective clinical management of the cancer patients, the early diagnosis and prognosis is the need of the hour. The advancing technologies in the field of medicine have led to the availability of enormous cancer data to the researchers. However, they have faced a prime challenge of predicting the accurate outcomes of this disease. For this very purpose, the concept of machine learning (ML) came into existence. The ML tools and techniques can detect key features from the electronic complex datasets to model cancer risks or patient outcomes. The ML techniques have been frequently used by the researchers in the field of bioinformatics and biomedicine to classify the cancer patients into low- and high-risk groups. Moreover, there has been an implementation of these techniques in modeling the progression of the tumor along with its effective treatment. Although the different ML algorithms have provided an improved understanding of the tumor condition, a proper validation is required for their applications in everyday clinical practices. Keeping in mind the growing trend of application of ML tools in cancer research, we present here a performance analysis of the three common classifiers: *artificial neural networks (ANNs), Naive Bayes,* and *support vector machines (SVMs)*. In this study, the effectiveness and accuracy of these classifiers have been compared in terms of their sensitivity, specificity, accuracy, and area under the curve (AUC) using Orange and R programming on the three different cancer sample datasets (viz., liver, prostate, and breast cancer).

Keywords

Cancer · Machine learning · ML algorithms · Artificial neural networks (ANN) · Naive Bayes · Support vector machine (SVM)

A. Saxena, S. Chandra, *Artificial Intelligence and Machine Learning in Healthcare*,
https://doi.org/10.1007/978-981-16-0811-7_8

8.1 Introduction

Machine learning (ML) has emerged with the innovations in the field of data sciences as a tool for automated classification. ML comprises a class of techniques and areas of research that can mimic the learning capacity of humans and enable the computer to learn and extract/classify patterns. ML is used in a broad range of applications from forecasting stock market regression to reinforcement learning to play games, but here we focus on prediction in the sector of healthcare.

The history of the relation between biological science and the field of machine learning is not new, but it is significant. The applications of ML methods using biological data are being used for the prediction of genes within and among species, functional annotation, and system biology and in the analysis of metabolic pathways. ML approaches are now being applied in medical science for the detection and classification of different types of tumors.

8.1.1 Machine Learning in Healthcare

From the past few decades, healthcare has become one such industry where digital data has exponentially increased. These data repositories are rich sources of divergent and interesting patterns. Since statistical techniques fall short in analyzing and extracting these patterns, ML techniques have been evolved to overcome them, and ML algorithms have a capability to extract, enfold, and transform the patterns from healthcare data.

The concept of introducing technology to the field of medicine started as a tool known as *expert systems*. ML are artificial intelligence (AI)-based techniques that comprise a variety of algorithms with a capacity to learn from situations and environment. These algorithms can build models for autonomous prediction and classification (Ahuja 2019). Similarly, in the past decade, the accuracy rate of cancer prognosis and diagnosis has not been up to the mark. So, in order to enhance the decision-making capability of clinicians, the elements of machine learning could be used to improve the accuracy levels in the pathology system (Sayed 2018). ML is able to find classes of algorithms that can show a high level of generalization performance to new datasets to which the machine (computer) is not exposed during the training.

Past studies were focused on designing of models to predict possible outcomes of a disease. These models were crafted for classification and prediction using supervised methods. Result analysis of such studies clearly necessitates the integration of multidimensional heterogeneous data; feature selection and classification techniques are favorable tools for cancer prognosis (Kourou et al. 2015). Digitalization in the field of clinical data handling and simultaneous popularity of deep neural networks (DNN) is another reason for the inclination toward ML in healthcare.

ML can detect patterns of certain diseases within patient's electronic healthcare records and notify clinicians about any anomalies. Artificial intelligence (AI) in medical studies usually performs clinical diagnoses and provides suggestions for the treatments using AI algorithms. AI has the power to deduce meaningful relationships within huge structured and unstructured datasets. Due to this capability, AI has been deployed in many clinical situations to diagnose, treat, and predict the possible outcome of a disease. Nowadays, machine learning, which is a subset of AI, plays a key role in many health-related applications, including the development of new medical diagnostics, management of patient's e-records, a doctor's prescription, and the treatment history. Medical diagnostic is a class of medical tests designed to detect infections and disease conditions in patients. ML-based diagnostics have been used to diagnose diseases by integrating cognitive computing with genome-based sequences. It helps patients to monitor health status so that he/she can maintain a healthy life. These medical diagnostics can be purchased by patients or may be used in laboratories. Additionally, AI has increased the ability of health professionals in deeply understanding the patterns of symptoms in patients and their response to treatment. Such AI systems have provided better feedback, guidance, and support in treating a patient, even in critical conditions (Delen et al. 2005).

8.1.2 Cancer Study Using ML

Cancer is the second leading cause of death at a global scale. Every year, a large population suffers due to this deadly disease. In 2018 itself, 9.6 million people have lost their lives because of cancer. Researchers from different domains are discovering the approaches to counter this disease. Early diagnosis of cancer helps in saving the life of many. ML-based cancer diagnosis provides better understanding of disease and preconditions with the level of severity of conditions of patients (Zhu et al. 2020). This chapter has been devised to describe, compare, and evaluate the performance of different machine learning techniques for cancer prediction and prognosis. Specifically the chapter discusses the various machine learning methods, the types of cancers along with their datasets, and the overall performance of these ML in early diagnosis and prognosis of cancer.

The literature reveals that a vast number of studies had worked upon the survival prediction problem using statistical approaches and artificial neural networks (ANN). However, very few studies related to medical diagnosis using decision trees (DT) had been conducted. Delen et al. used ANN, DT, and logistic regression (LR) to design prediction models for breast cancer survival (Delen et al. 2005). Lundin et al. used ANN and logistic regression models for cancer survival prediction using 5-, 10-, and 15-year breast cancer repositories (Lundin et al. 1999). Pendharker et al. also used several data mining techniques to unveil new patterns in breast cancer. The study used the ML algorithms in establishing similarities between the cancer cases and, thus, helped in proper determination and treatment of cancer (Pendharkar et al. 1999). These studies are only a few examples of research that utilize ML to medical fields for prediction of diseases. The recurrence of breast cancer can also be determined using various data mining techniques or ML tools.

While the artificial neural networks predominates, a number of other ML strategies are being used in cancer prediction. The overall performance and predictive accuracy of cancer prognosis process is highly improved with incorporation of ML tools (Cruz and Wishart 2007).

8.2 Machine Learning Techniques

The aim of ML algorithms remains to develop a mathematical model that fits the data (Mitchell 2006). ML study basically deals with the creation and evaluation of algorithms that facilitate pattern recognition, classification, and prediction, based on the models derived from existing data. ML algorithm consists of a learning attribute which is needed to acquire knowledge from data. There are two types of ML-based problems, one where prior knowledge about data is needed and one where prior knowledge is not required. Problems based on prior knowledge requirement are named as prediction and classification. ML designs model for such problems.

During model construction data are supplied to algorithms for learning purpose. Firstly, ML algorithm is trained from data to constitute a model, and then another set of data is provided to test the model. In this process, ML uses a learning coefficient to learn from train data and apply that knowledge on test data. In other problems, algorithms are directly applied to the final dataset without any training, and algorithms do not learn anything from data. Based on this learning attribute, ML algorithms have majorly categorized into four types as shown in Fig. 8.1.

In *supervised learning* the class membership of new objects is accurately predicted on the basis of previously available features, whereas in unsupervised learning there are no such predefined labels for the objects. In such a case, unsupervised algorithms such as clustering explore the data to infer similarities between objects. The similarities are helpful in defining groups of objects called clusters. Similarly association algorithms such as Apriori and FP are used to devise association between unlabeled dataset. The natural groupings in the data are easily identified in unsupervised algorithms. Thus, the two learning approaches are quite opposite to each other. In supervised learning, the data come up with class labels, and algorithms classify the classes with labeled data; in unsupervised learning, data are completely unlabeled, and the learning methods works in defining the labels and classifying objects with them. Semi-supervised algorithms follow a blended approach of supervised and unsupervised learning. During the training, the amount of unlabeled data exceeds the labeled data. Self-learning algorithms, also known as reinforcement learning, are capable enough to learn by themselves. They gradually learn from labeled data and drive their own logics and understanding. Figure 8.2 shows different ML algorithms based on their learning quotient and application.

Generally there are four types of analysis popular in medical diagnostics: descriptive, inferential, predictive, and prescriptive. Machine learning is usually employed for inferential, predictive, and prescriptive analysis. These algorithms perform five types of tasks to provide such analysis. Tasks are association, clustering, dimension reduction, classification, and prediction. The performance of each task and algorithm is measured by number of metrics. Figure 8.3 illustrates the tasks and metrics. These

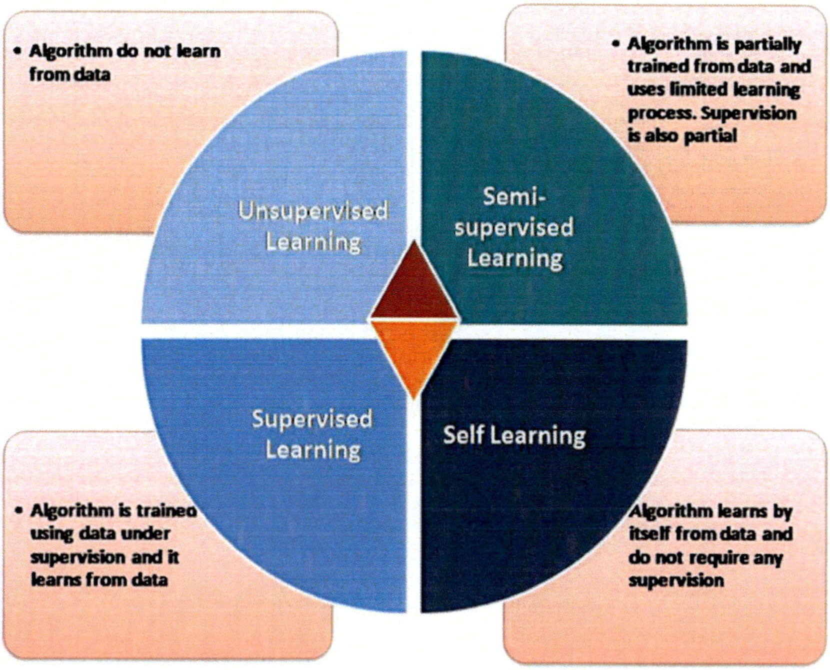

Fig. 8.1 Categorization of machine learning algorithms

metrics are tools to measure the progress of algorithms. They guide researchers to finalize and verify the ML-based model for their study (Fortunato et al. 2019).

Classification and prediction are generally used for designing the forecasting model. In classification algorithms, the data item is classified into one of the predefined class using the learning function. In prediction, the learning function predicts the value. When a classification or prediction model is developed using ML techniques, training, testing, and classification or prediction errors are produced. The former ones are the misclassification errors on the training data, while the latter ones are the expected errors on testing data. It should be noted that a high-quality model always fits the training set well and also correctly classifies all the instances. If the test error rate of a model is high even though the training error rates are low, then the model suffers with overfitting issues. Overfitting of the model occurs when data is unbalanced or algorithm is not suitable for the dataset. While designing a model, factors such as unbalanced class, overfitting, and underfitting are always essentially considered. These factors influence the accuracy and precision of a model. Majorly health care studies employ prediction, dimension reduction, and classification.

The performance of these models can be elevated by taking the right decision on class balancing and algorithm selection. Metrics are defined to register and observe the performance of these ML algorithms. Researchers can easily validate their results

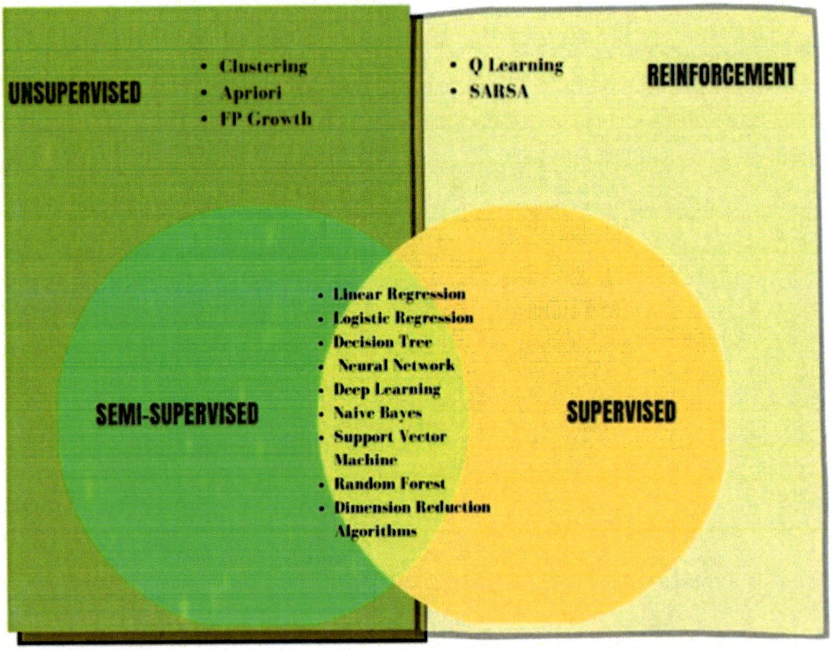

Fig. 8.2 Machine learning algorithms

and derive conclusions using these metrics. Many studies have shown usage of metrics in cancer prediction assessment (Fortunato et al. 2019).

8.3 Machine Learning and Cancer Prediction/Prognosis

8.3.1 Cancer: The Dreaded Disease and a Case Study for ML

In recent years, machine learning algorithms are being used for analysis of the occurrence of disease in several cancer patients. Machine learning provides a more graceful and effective solution for the classification of cancer patients. The symptoms of cancer may be misleading, and, in turn, it delays the process of diagnosis and treatment. Over 100 types of cancers affect humans (Hanahan and Weinberg 2011). Figure 8.4 shows various applications of ML in cancer study.

For this article we have selected the three most important types of cancers for the purpose of ML.

Liver Cancer

Hepatocellular carcinoma (HCC) is considered as the prime liver malignancy. It has even contributed predominantly in the cancer-related deaths worldwide (Crissien and Frenette 2014). This is also known as primary liver cancer. Other types of liver

Task		Metrics
Association Rule	Algorithms find the relationships between the given data items	Support and confidence
Clustering	Algorithms split the datasets in different clusters using a mathematical formula.	•Internal: Silhouette index, Dunn index etc •External :Purity ,Entropy etc •Relative: Mix of above
Dimension Reduction	Algorithms are used for reducing the dimensional complexity of data	Local preservation criteria and Global preservation criteria
Classification	Algorithms classify a dataset based on the class of output variable/s	Confusion matrix,Accuracy, Precision and ROC
Prediction	Algorithms are used to forecast the possibility of object or event	Accuracy, Precision and RMSE

Fig. 8.3 Tasks and metrics

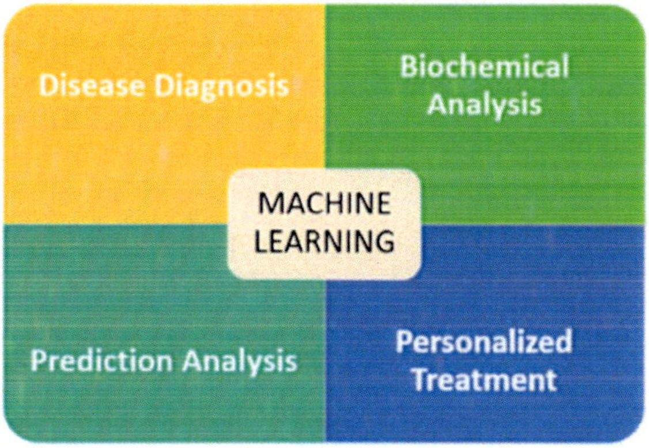

Fig. 8.4 Applications of ML in cancer prediction/prognosis

cancer, such as intrahepatic cholangiocarcinoma and hepatoblastoma, are much less common. Cancer that starts in another organ, such as the colon, breast, or lung, and then spreads to the liver is called secondary liver cancer. Secondary liver cancer is more common than primary liver cancer (Ferlay et al. 2010). The only hope for effective treatment of liver cancer is early and correct detection. Several biochemical markers can be effectively used for this, which in turn predisposes the disease in patients.

Prostate Cancer
Prostate is a small walnut-shaped gland in the male reproductive system responsible for the production of the seminal fluid that nourishes and transports sperm. Cancer in this gland is one of the most common types of cancer in males. Most of the prostate cancer grows slowly and remains confined to the prostate gland only; however, some types of aggressive prostate cancer can spread quickly from the prostate to other areas of the body, particularly the bones and the lymph nodes. Though asymptomatic initially, in later stages, it can lead to difficulty in urinating, blood in the urine, or pain in the pelvis and back. Factors that increase the risk of prostate cancer include older age, a family history of the disease, and race. The screening of some biochemical parameters can lead to the development of reliable and specific diagnosis of prostate cancer, which if detected earlier has a better chance of successful treatment (Barlow et al. 2019).

Breast Cancer
Breast cancer remains the most common invasive cancer among women. It is a multistep process concerning multiple cell types. Substantial increase in breast cancer consciousness and research funding has helped to create advances in the early detection and treatment of breast cancer. In the United States, breast cancer remains the major cancer among women with approximately 190,000 new cases annually. Breast cancer incidence rates have increased, but the number of deaths linked with this disease is progressively declining, largely due to factors such as earlier detection, a new modified approach to treatment, and a better understanding of the disease (Baralt and McCormick 2010).

8.3.2 Machine Learning in Cancer

Cancer detection at an early stage has become essential in cancer research for better clinical management of patients. In this study, three different datasets of liver, prostate, and breast cancer were examined to screen the budding cancer at an early stage long before the development of any visible symptoms. Because of the advancement in computational technology and medical science, huge data repositories of cancer data are generated for clinical research. The bank of datasets is easily available for cancer analysis and related research (Nagy et al. 2020). However, the most interesting and challenging task in cancer study is the accurate prediction of a disease outcome. Various ML techniques can be used for the classification of cancer patients. Different classifiers were used for the analysis purpose, among them are *artificial neural networks (ANN)*, *Naive Bayes classifier*, and *support vector machine (SVM)* which provide effective models with highly accurate results. Problems such as determining the class of cancer patients based on their risk (as high or low), treatment susceptibility, and cancer growth with impacts are analyzed using ML methods. The ML tools and techniques are being used to predict the precise progression state and a valid treatment of the cancers. Even though ML methods have improved the understanding of cancer progression, an appropriate level of validation is the need of the hour so that these models are adapted easily in everyday clinical practices (Murali et al. 2020). In this chapter, the predictive models discussed are based on various supervised ML techniques as well as on different input features and data samples. Based on the analysis of results, it was found that these classifiers provide satisfactory performance in terms of accuracy, recall, precision, specificity, and other parameters. The overall prognosis of various types of cancer can be improved by the application of ML techniques (Obaid et al. 2018).

8.3.3 Dataset for Cancer Study

Machine learning algorithms require large amounts of raw datasets for data exploration, data mining, and statistical analysis. Data collection can be done from various sources after which it is processed to remove missing value and corrupt data detection imputation. In this chapter, the datasets have been taken from https://www.kaggle.com (an online community of data scientists and machine learners, owned by Google LLC). There are three different datasets about liver, prostate, and breast cancer, containing important variables such as occurrence of disease, gene expression, proteins, environmental factors and diagnosis, and survivability rates. The present study was focused to analyze cancer based on the following:

- Diagnosis of disease.
- Occurrence of cancer.
- Survivability rate.

Liver cancer dataset: This dataset contained 416 liver patient records and 167 non-liver patient records collected from North East of Andhra Pradesh, India. The "dataset" column is a class label used to divide groups into liver patient (liver disease = 1) or not (no disease = 2). This dataset contains 441 male patient records and 142 female patient records. Total records are 583.

Any patient whose age exceeded 89 is listed as being of age "90." Liver dataset has the following parameters segregated into covariables, factors, and dependent variable as shown in Table 8.1.

Prostate cancer dataset: There are 12 parameters which were divided into covariables and dependent variable as shown in Table 8.2.

Breast cancer dataset: Breast cancer dataset has nine parameters which were divided into covariables, factors, and dependent variable as shown in Table 8.3.

Table 8.1 Liver cancer dataset

Covariables	Factors	Dependent variable
Age of patient	Gender of patient	Occurrence of cancer: This variable gives the value of occurrence or nonoccurrence
Total bilirubin		
Direct bilirubin		
Alkaline phosphatase		
Alanine aminotransferase		
Aspartate aminotransferase		
Total proteins		
Albumin and globulin ratio (blood protein)		

Table 8.2 Prostate cancer dataset

Covariables	Dependent variable
Patient no.	Diagnosis
Patient ID	
Radius	
Texture	
Perimeter	
Area	
Smoothness	
Compactness	
Concavity	
Symmetry	
Fractal dimension	

Table 8.3 Breast cancer dataset

Covariables	Factors	Dependent variable
Patient subject ID	ER (hormone estrogen)	Survivability rate
Age	PR (hormone progesterone)	
MRI (magnetic resonance imaging)	HR (hormone receptor)	
	Bilateral	
	PCR (polymerase chain reaction)	

Covariables are those parameters whose value is decimal or integer. *Factors* are those variables whose value depicts some class or order of the class. The nature of parameters and dependent variables helps in deciding the type of ML model, i.e., classification or prediction. If the dependent variable is defined as a class, then classification model is designed. If dependent variable has decimal values, then prediction model is constructed.

8.3.4 Steps to Implement Machine Learning

Designing a model using machine learning algorithms is a step-by-step process also known as knowledge discovery process. Each step has a defined set of tasks. Major steps are problem definition with identifying features, data collection, data cleaning, model framing, and evaluation of the model as given in Fig. 8.5. Researchers can frame their methodology using these steps.

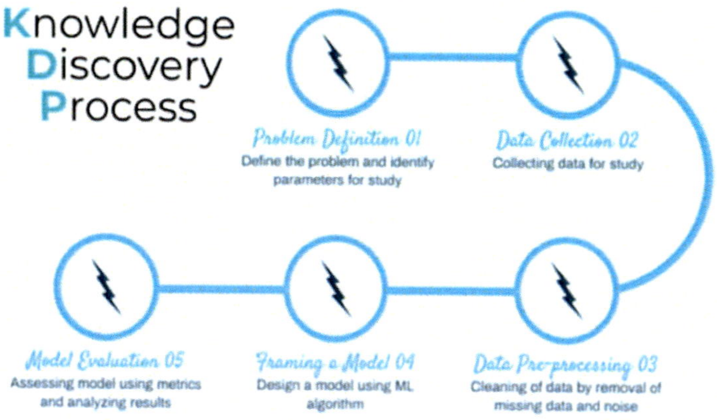

Fig. 8.5 Knowledge discovery process

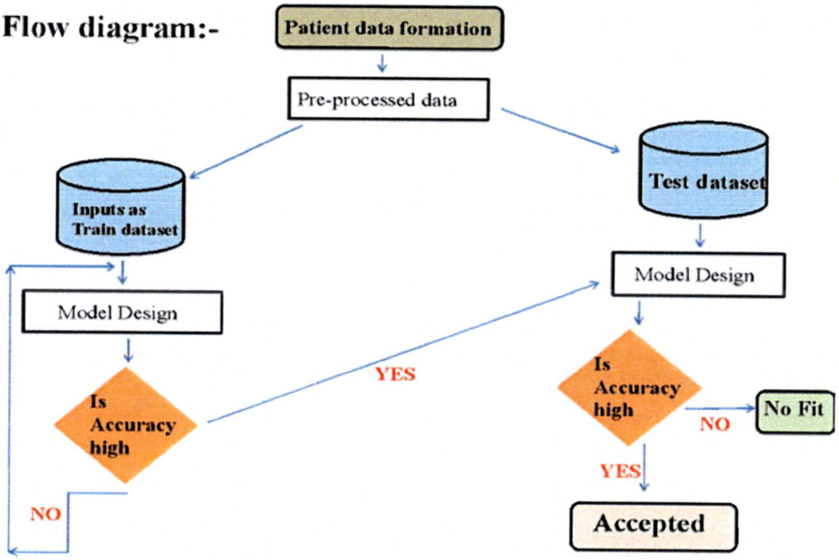

Fig. 8.6 Flowchart for cancer prediction using ML

Figure 8.6 shows a flowchart for cancer prediction using ML in our case study. After the collection and preprocessing of data, model is framed. If the data is unlabeled, unsupervised learning algorithm is used for model framing. Labeled data is the dataset where outcomes are already known. If data is properly labeled, then dataset is divided into two parts: train and test set, and then model is designed using semi-supervised or supervised learning algorithms. In this study, we have labeled datasets. Cross-validation is another important step used for validation of results. In such case data can be divided into three sets instead of two: train, test, and validation set, respectively.

8.3.5 Tool Selection for Cancer Predictions

There are different tools in the market for ML designing. Python and R are popular languages for ML coding. The tool selection depends upon a user's understanding of coding and comfort of usage. There are some tools that provide a smooth interface to apply ML in any dataset. These tools are very useful in the initial stage of research. They help researchers gain basic understanding of ML process without the burden of coding. Also, they provide clear and effective visualization of results. In this chapter, Orange tool and RStudio have been discussed. Research needs to learn R programming to use RStudio. In the case of Orange, users can simply drag and design their model.

Orange

Orange is an open source tool with component-based visualization facilities for data visualization, machine learning, and data analysis. It is a data mining tool frequently used in biomedicine, bioinformatics, genomic research, and teaching. It also provides a python coding interface for python library plug-in. Orange components are termed as widgets that include simple data visualization, subset selection and preprocessing, and empirical evaluation of learning algorithms and modeling. Workflows can be created by linking predefined or user-designed widgets. Widgets can also be edited using python coding.

RStudio

RStudio is an integrated development environment (IDE) for R. R is an easy but extensive programming language for data manipulation, analysis, and visualization. It is popular among data scientists for its effective data handling and storage facilities. RStudio includes a console, syntax-highlighting editor that supports direct code execution, as well as tools for plotting, history, debugging, and workspace management. A GNU package, source code for the R software environment, is written primarily in C. It includes a large number of libraries to support medical and bioinformatics analysis.

8.3.6 Methodology, Selection of ML Algorithm, and Metrics for Performance Measurement of ML in Cancer Prognosis

Methodology

The first step requires data collection. In this study, we have collected a dataset from an online data repository. Dataset comprises population characteristics, age, gender, and factors responsible for predicting survival rate. Dataset involving breast, liver, and prostate cancer was considered. The final dataset for the cancer patients was developed after the implementation of data cleansing and preprocessing strategies. The dataset is then split into two sets: train and test. In the next stage, an algorithm was selected after reviewing literature, and then the model is framed from train and test data. Figure 8.6 describes the methodology flowchart for the current study.

The current study of cancer has three predicted purposes: (1) the evaluation risk assessment or susceptibility of the cancer under observation, (2) the evaluation of occurrence/reoccurrence or local control of the developing cancer, and (3) the evaluation of the survival rates. In the first two cases, there is a chance of developing or redeveloping a type of cancer after complete or partial remission. In the last case, the prime objective is to predict the disease-specific survival rate or the overall survival rate after the development of cancer. The prediction of cancer patient outcome usually deals with life expectancy, survivability, progression, treatment, and diagnosis. For this purpose, Orange and R programming are used. These algorithms are used to extract instances from large datasets, to create statistical software, graphics, and data analysis. Both these tools are also an open source data mining tool and allow a user to perform ample of data mining algorithms, which involve collection of tools for data classification, regression, clustering, association rules, and visualization.

Selection of Machine Learning Algorithm for Cancer Study

There are a vast number of ML algorithms for cancer prediction. According to the literature, support vector machine (SVM), neural networks (NN), and Naive Bayes provide high accuracy and better precision. In this chapter all the three machine learning algorithms are discussed.

Support Vector Machine (SVM)

SVM is used in the present study as it is an emerging powerful machine learning technique and one of the most utilized methods for breast cancer diagnosis. The term SVM was first suggested by Vapnik on the foundation of statistical learning theory. It is mainly created for classification analysis. It is also used to classify both linear and nonlinear data. The main advantage of this classifier is to discover the improved decision border, which examines the greatest decisiveness (maximum margin) among the classes. SVM has also been used previously in the field of bioinformatics as a promising tool for pattern recognition, cancer prognosis, and diagnosis. Figure 8.7 shows the SVM model with five output classes.

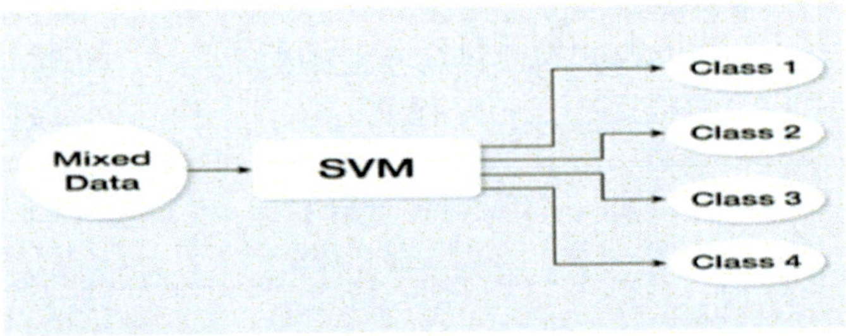

Fig. 8.7 SVM with different classifiers. Source: https://miro.medium.com/max/2560/1*dh0lzq0QNCOyRlX1Ot4Vow.jpeg

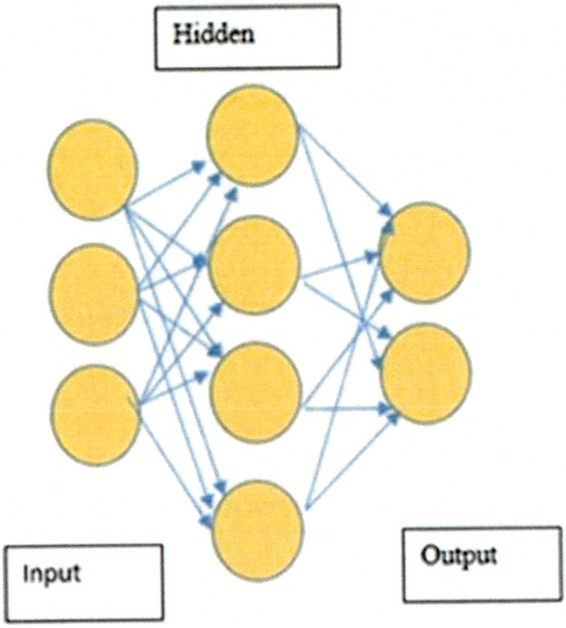

Fig. 8.8 An example of artificial neural networks

Artificial Neural Networks (ANN)

Neural networks is a model for receiving, processing, and transmitting information in terms of computer science. A set of input data is mapped into an appropriate set of output data using various features of the multilayer perceptron (MLP) model. The neurons in the input layers play a specific role in dividing the input signal between neurons in the hidden layer. An identical fashion is followed for the determination of the output of neurons in the output layer. For these types of classification problems, MLP is said to perform better than other available ANN architectures (Obafemi et al. 2019). Figure 8.8 provides ANN structure.

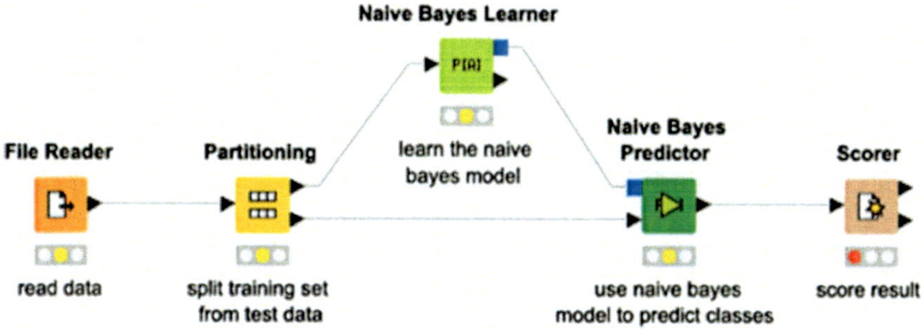

Fig. 8.9 The flow diagram of Naive Bayes in machine learning (Source: https://i.stack.imgur.com)

Naive Bayes

Naive Bayes is another ML algorithm for classification problems. It is working based on Bayes' probability theorem. It is used to resolve problems associated with text and web classification, which deals with HD training datasets. It is the first algorithm that is designed to resolve text classification problems. Figure 8.9 is a flow graph of Naive Bayes.

Metrics for Performance

Orange and R language tools were used to determine the model with high accuracy and better performance. The following metrics were used to assess the performance:

- Train time = More time the model takes; it will predict the best accuracy.
- AUC (area under the curve) = A measure of how well a parameter can distinguish between occurrence and nonoccurrence of cancer. The higher the AUC, the better the model is predicting. It gives an overall performance of a classification model. The area under the ROC curve depicts the measure of separability. The value of AUC ranges from 0 to 1. AUC below 0.5 shows a failed test model. A perfect test has an area of 1.00. It has zero false positives and zero false negatives. So, in order to yield correct results, the test should have an area between these two values. The area is reported as a fraction even if the results are plotted as percentages. Figure 8.10 shows the ROC curve.
- CA (cumulative accuracy) = It observes average accuracy of all the models.
- F1 = Harmonic mean of precision and recall.
- Sensitivity = The fraction of people with the disease that the test correctly identifies as positive. The formula is Sensitivity = True Positives/(True Positives + False Negatives).
- Specificity: The fraction of people without the disease that the test correctly identifies as negative. The formula is Specificity = True Negatives/(True Negatives + False Positives).
- ROC curve = The ROC (receiver operating characteristic) curve is a promising tool used to predict the probability of a binary outcome. The graph is plotted for a number of different candidate threshold values between 0.0 and 1.0 while keeping

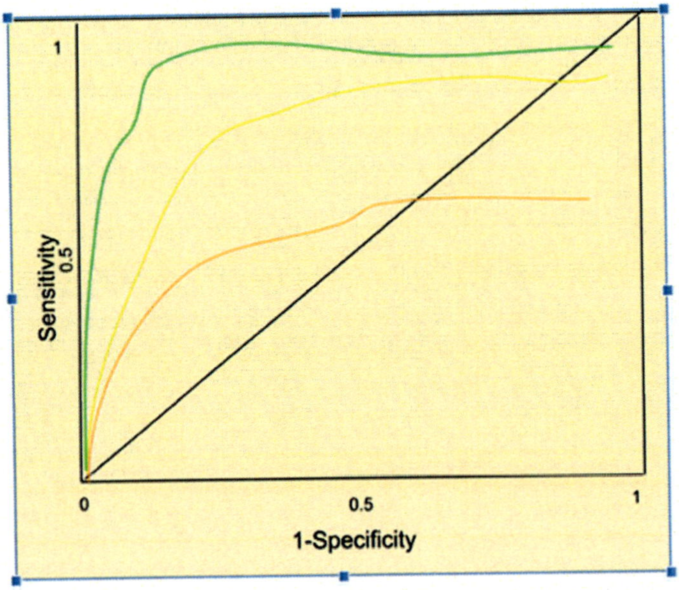

Fig. 8.10 ROC curve

the false positive rate on the x-axis and the true positive rate on the y-axis. The prime utilization of the ROC curve is in deciding where to draw the line between "normal" and "not normal." The decision will become easier if all the control values exceed or are lower than all the patient values. However, in reality, these two distributions overlap, and hence the decision-making process is not that easy. If the threshold value is increased, those who do not have the disease would not be mistakenly diagnosed, but there are chances that some of the diseased people are missed. If the threshold value is lowered, there will be a correct identification of almost all diseased people, but at the same time there are chances of diagnosing the disease in more people than the actual ones.

The aim of the study was to get the highest *accuracy* and *specificity* for the various classifiers. Furthermore, the accuracy of the three classifiers is compared in order to recognize which classifier works better for the classification of different kinds of cancer. All classifiers with their types are rated based on these standards, the overall accuracy, the specificity, and the time taken to construct the model.

8.4 Results and Analysis

8.4.1 Liver Cancer Dataset

Patients with liver cancer have been continuously increasing. The prime reasons behind this could be periodic and excessive consumption of alcohol, exposure to harmful gases, and consumption of contaminated food, pickles, and drugs. The prediction algorithm was evaluated using the liver cancer dataset, reducing the overall burden on the doctors. Our aim was to use this patient record to determine which patients have developed liver cancer and which do not. Figure 8.11 shows the flowchart in Orange tool for designing ML models.

Figure 8.12 shows the comparative table of SVM, NN, and Naive Bayes for liver cancer data.

The model was constructed using tenfold cross-validation with 50% training dataset. The performance of each model is compared using the area under the curve (AUC); cumulative accuracy (CA); F1 score, which is a weighted harmonic mean of precision and recall; precision; and recall. According to the results, NN has the highest CA, i.e., 78%, recall (78%), and precision (77%) with F1 score of 76%. F1 score is good when it is near to 100%. It refers model performance in combination of precision and recall. Overall NN is constructing a better model.

Figures 8.13a, 8.13b, and 8.13c shows the confusion matrices of all the three algorithms. The dataset size was 583, out of which 167 cases belong to class 2 and 416 cases belong to class 1.

Figure 8.13a shows SVM confusion matrix for liver dataset.

The results are as follows:

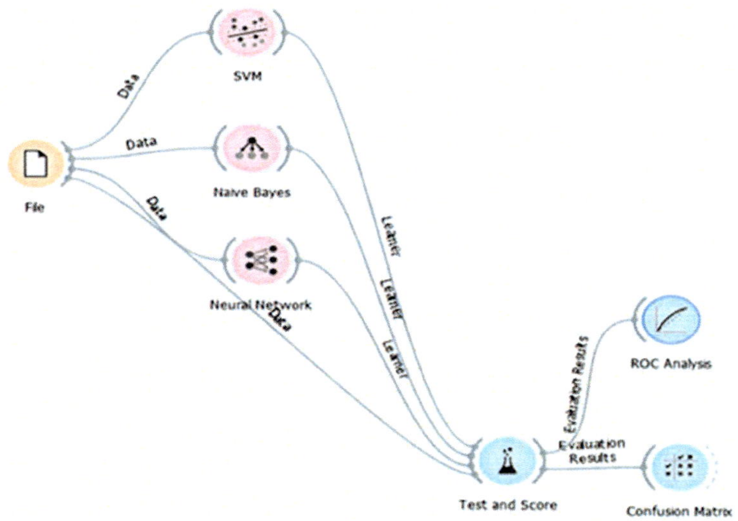

Fig. 8.11 Flowchart in Orange tool

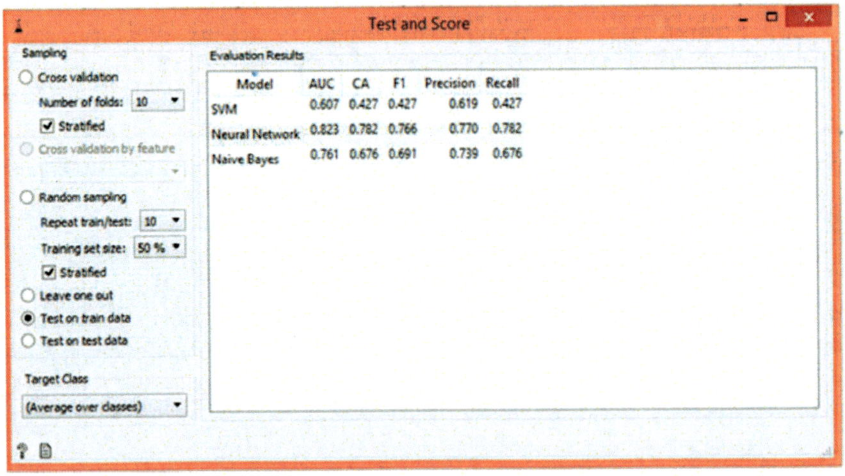

Fig. 8.12 Performance comparison of machine learning models

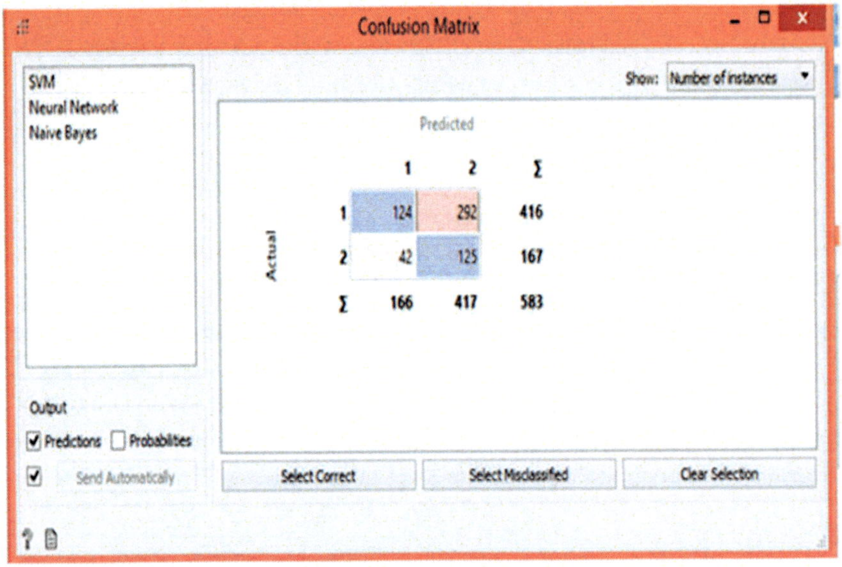

Fig. 8.13a Confusion matrix for liver cancer dataset using SVM

- There are 124 cases which actually belong to class 1 and predicted also 1.
- There are 292 cases which actually belong to class 1 but predicted as 2.
- There are 42 cases which actually belong to class 2 but predicted as 1.
- There are 125 cases which actually belong to class 2 and correctly predicted as 2.

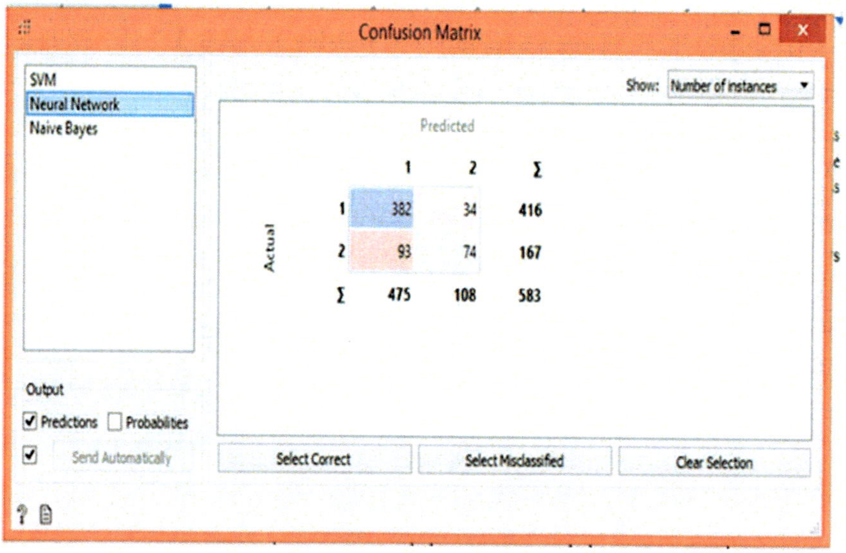

Fig. 8.13b Confusion matrix for liver cancer dataset using NN

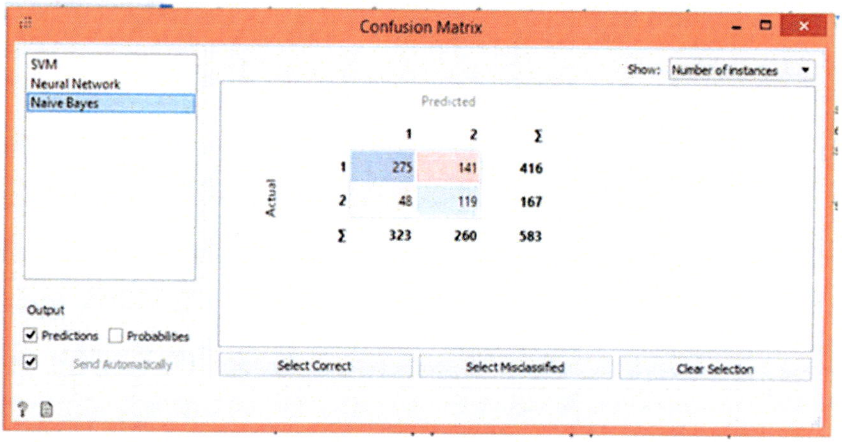

Fig. 8.13c Confusion matrix for liver cancer dataset using Naive Bayes

Therefore, out of 416 liver patients, SVM is able to predict only 124 cases, and out of 167 non-liver patients, SVM is able to predict only 125 cases accurately.

Figure 8.13b provides neural networks confusion matrix for liver dataset.

The following are the results:

- There are 382 cases which actually belong to class 1 and predicted also 1.
- There are 34 cases which actually belong to class 1 but predicted as 2.

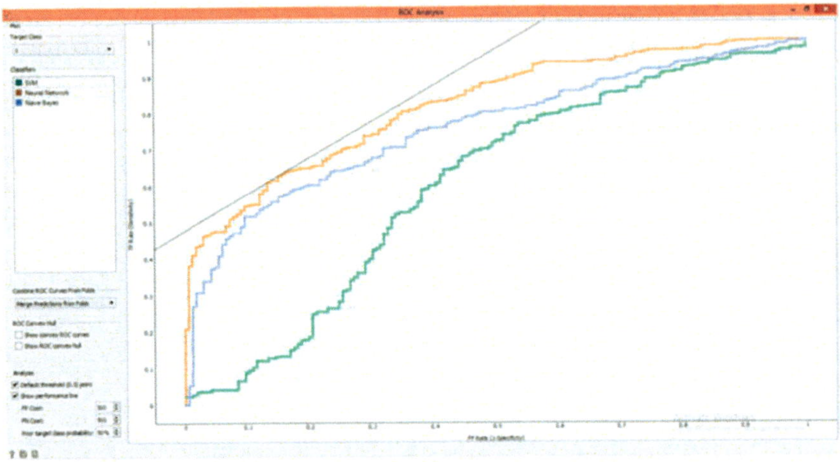

Fig. 8.14a ROC curve for class 1

- There are 93 cases which actually belong to class 2 but predicted as 1.
- There are 74 cases which actually belong to class 2 and correctly predicted as 2.

Hence, out of 416 liver patients, NN is able to predict only 382 cases, and out of 167 non-liver patients, NN is able to predict only 74 cases accurately.

Figure 8.13c illustrates Naive Bayes confusion matrix for liver dataset.

The results are as follows:

- There are 275 cases which actually belong to class 1 and predicted also 1.
- There are 141 cases which actually belong to class 1 but predicted as 2.
- There are 48 cases which actually belong to class 2 but predicted as 1.
- There are 119 cases which actually belong to class 2 and correctly predicted as 2.

Thus, out of 416 liver patients, Naive Bayes is able to predict only 275 cases, and out of 167 non-liver patients, Naive Bayes is able to predict only 48 cases accurately.

From the confusion matrices, it could be concluded that NN is able to predict liver patients more precisely, whereas non-liver patients are predicted better by SVM.

The trade-off between the true positive rate and false positive rate is summarized using the ROC curves for a predictive model using different probability thresholds. The quantification of the overall ability of the test to discriminate between the diseased individuals and those without the disease is done by determining the area under the ROC curve. Figures 8.14a and 8.14b shows the ROC of all the three models.

AUC is referred further for inference. The TPR and FPR for every possible threshold value of the classifier are obtained, and then the graph is plotted between 0 and 1. A perfect test has an area of 1.00. In the given model AUC are:

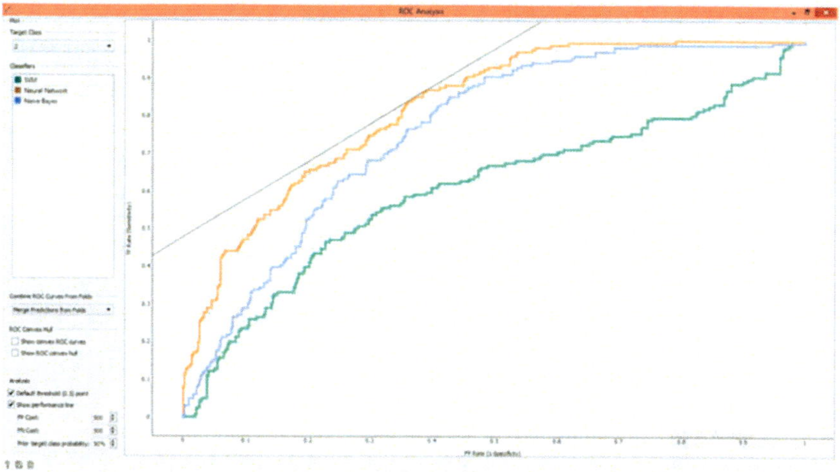

Fig. 8.14b ROC curve for class 2

- AUC of SVM is 0.607.
- AUC of Naive Bayes is 0.761.
- AUC of NN is 0.823.

Inference The AUC of NN is high, so it is sufficient and clinically acceptable. It shows that NN performs the best.

Neural networks (NN) is the best prediction model for the liver cancer dataset. Further, to cross-examine the results, we executed the same model in RStudio. This shows three layers of neurons: an input layer where the independent variables or inputs of the model are accepted, followed by one hidden layer, and an output layer where final predictions are generated. Figure 8.15 shows an NN model designed in R.

This is the three-layer neural networks model, which involves an inner layer containing independent parameters, a middle layer which has been processed, and an output layer which is also called diagnostic output containing dependent variables.

Commands for R

```
>nn.results<- compute(nn, temp_test)
>results<- data.frame(actual = testset$Disease, prediction
=nn.results$net.result)
>roundedresults<-sapply(results,round,digits=0)
>roundedresultsdf =data.frame(roundedresults)
>attach(roundedresultsdf)
>table(actual,prediction)
>prediction
```

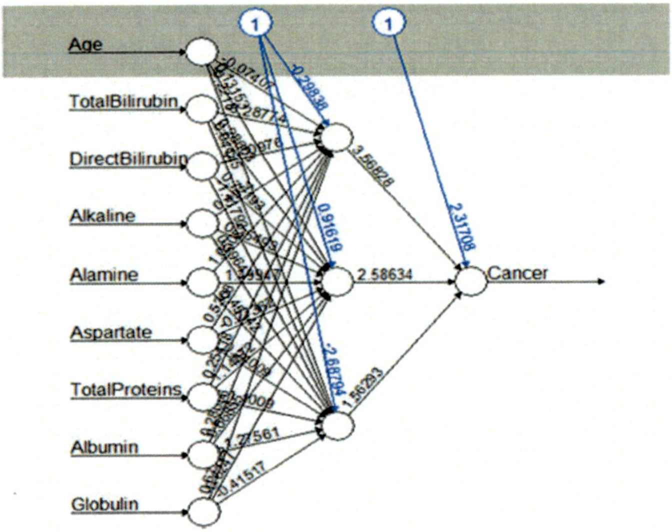

Fig. 8.15 Neural networks model using RStudio

Table 8.4 Confusion matrix generated by ANN for liver cancer dataset in RStudio

	Prediction	
Actual	0	1
0	396	20
1	82	85

The dataset was divided into 20–80 for test/train. The R commands produced confusion matrix. Here 0 is the liver cancer patients, and 1 is the case without liver cancer as shown in Table 8.4.

In this, NN performance has increased as 396 cases are correctly discovered out of 416 infected cases. Similarly 82 cases are correctly identified out of 168 as noninfected cases.

8.4.2 Prostate Cancer Dataset

Our dataset includes back pain symptoms that are classified as abnormal or normal. Prostate cancer is a disease in which malignant (cancer) cells form in the tissues of the prostate. Our dataset contains prostate cells that are classified as good or bad and are appropriate for the use of predictive models. This dataset includes 12 variables. Our aim was to use this patient record to determine which patients have diagnosed with disease and which do not. Orange and R language tools were used to determine the best accuracy model and better performance. Figure 8.16 illustrates the performance of all the models on the prostate cancer dataset.

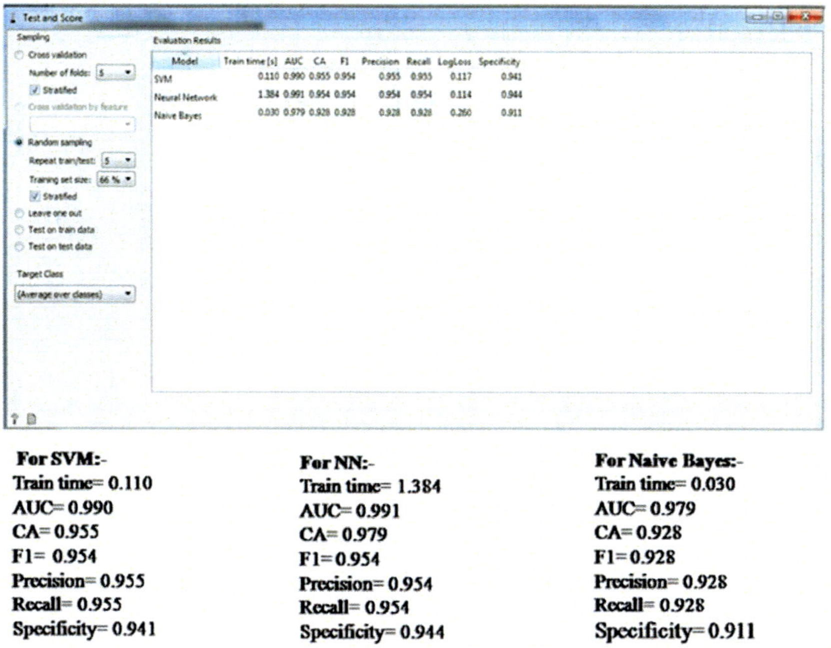

Fig. 8.16 Predictive model using the Orange tool on prostate cancer dataset

After a thorough analysis of the results of the different classifier models, these results were compared with each other. From Fig. 8.16, it can be concluded that NN alone performed the best in terms of specificity (94.4%), precision (95.4%), recall (95.4%), and accuracy (94.4%). SVM is also equally good and better than Naive Bayes.

Figure 8.17a shows SVM confusion matrix for the prostate cancer dataset.

The results are as follows:

- There are 594 cases which actually belong to class B and predicted also B.
- There are 16 cases which actually belong to class B but predicted as M.
- There are 28 cases which actually belong to class M but predicted as B.
- There are 332 cases which actually belong to class M and correctly predicted as M.

Therefore, out of 610 patients with benign cancer, SVM is able to predict only 594 cases, and out of 360 with malignant cancer, SVM is able to predict only 332 cases accurately.

Figure 8.17b shows Naive Bayes confusion matrix for the prostate cancer dataset.

The results are as follows:

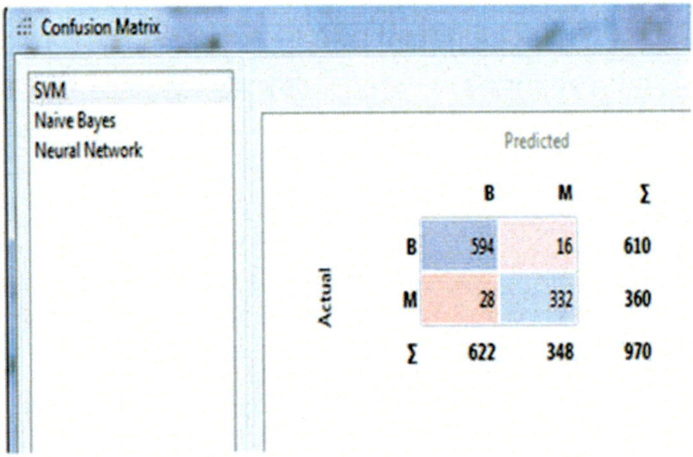

Fig. 8.17a Confusion matrix for prostate cancer dataset using SVM

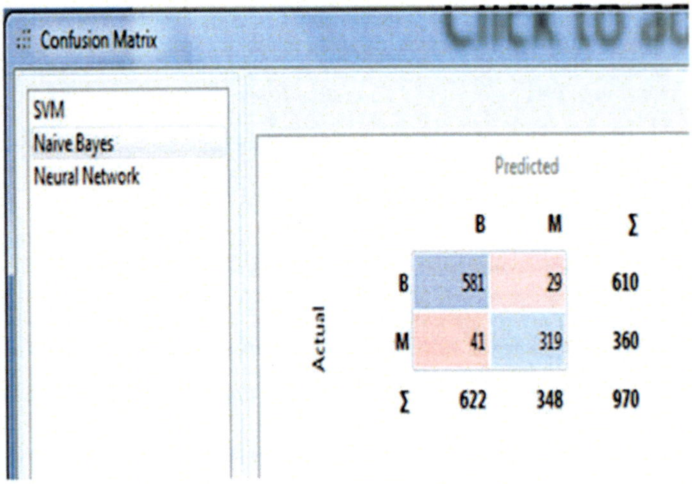

Fig. 8.17b Confusion matrix for prostate cancer dataset using Naive Bayes

- There are 581 cases which actually belong to class B and predicted also B.
- There are 29 cases which actually belong to class B but predicted as M.
- There are 41 cases which actually belong to class M but predicted as B.
- There are 319 cases which actually belong to class M and correctly predicted as M.

Therefore, out of 610 patients with benign tumor, Naive Bayes is able to predict only 581 cases, and out of 360 with malignant cancer, Naive Bayes is able to predict only 319 cases accurately.

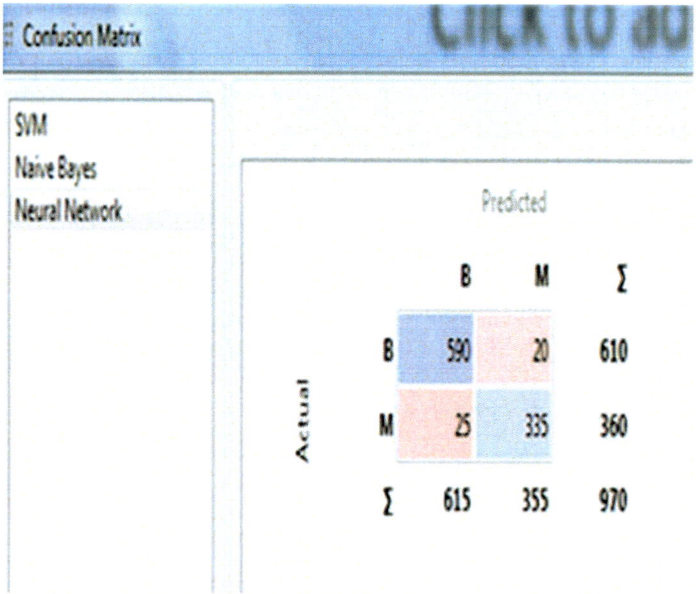

Fig. 8.17c Confusion matrix for prostate cancer dataset using neural networks

Figure 8.17c shows NN confusion matrix for the prostate cancer dataset.
The results are as follows:

- There are 590 cases which actually belong to class B and predicted also B.
- There are 20 cases which actually belong to class B but predicted as M.
- There are 25 cases which actually belong to class M but predicted as B.
- There are 335 cases which actually belong to class M and correctly predicted as M.

Therefore, out of 610 patients with benign tumor, NN is able to predict only 590 cases, and out of 360 with malignant cancer, NN is able to predict only 335 cases accurately.

Figure 8.18 illustrates the ROC curve for prostate cancer.

The curve shows the receiver operating characteristics and summarizes the trade-off between the true positive and false positive rates for a predictive model:

- AUC of Naive Bayes is 0.97.
- AUC of SVM is 0.99.
- AUC of NN is 0.99.

The datasets were trained using the three methods: ANN, SVM, and Naive Bayes, followed by the development of the classifier model. After this, different performance metrics were used to observe their respective results. The maximum value of

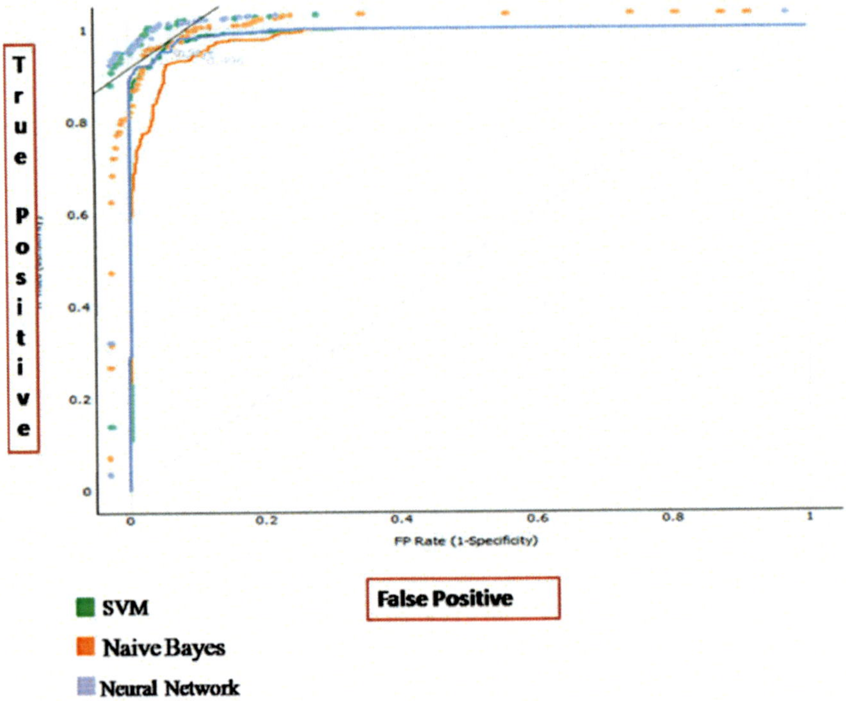

Fig. 8.18 Curve of receiver operating characteristics for prostate cancer dataset

the area under the curve (AUC) was observed in the neural networks, that is, 0.990, as compared to Naive Bayes. Neural networks (NN) is the best prediction model for the prostate cancer dataset. The three layers of neurons are visible here. The AUC of NN is higher than other classification models; as a result, NN is sufficient for clinical diagnosis.

Figure 8.19 shows the three-layer neural networks, which involves an inner layer containing independent parameters, a middle layer which has been processed, and an output layer which is also called diagnostic output containing dependent variables.

In Fig. 8.20, NN performance has been evaluated using 50–50 split and 70–30 split. This time the size of the test data in the 50–50 split was 284. In 285, 67 were of class B (shown as 0 in Fig. 8.20), and 218 were of class M (shown as 1 in Fig. 8.20). Briefly, the NN performed at 98% accuracy.

Similarly, in the case of 70–30 split, the size of the test dataset was 200, where 132 were of class M and 39 was of class B. NN accuracy was 99%.

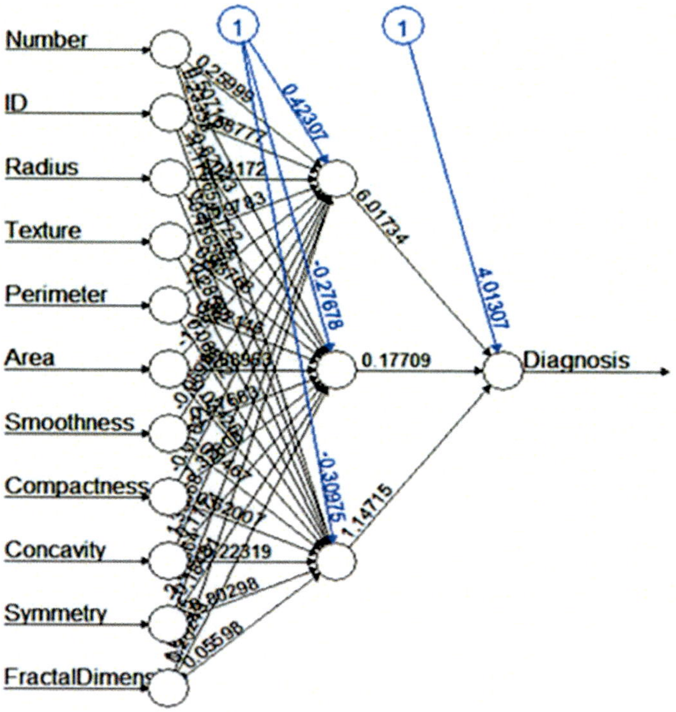

Fig. 8.19 Neural networks model by RStudio

- **For 50-50%,** Train=1-284 and Test=285-569

	prediction	
actual	**0**	**1**
0	**65**	**2**
1	**13**	**205**

- **For 70-30%,** Train=1-398 and Test=399-598

	prediction	
actual	**0**	**1**
0	**38**	**1**
1	**11**	**121**

Fig. 8.20 Classification matrix of neural networks model by RStudio

8.4.3 Breast Cancer Dataset

In breast cancer, malignant cell growth occurs in the breast. The cancer has a potential to spread to other parts of the body, if left untreated. This cancer is the most common type of cancer in women of the United States. In every three cancers diagnosed, one case is of breast cancer. Our aim was to use this patient record data to determine the survivability rate of patients suffering from breast cancer. Orange and R language tools were used to determine the best accuracy model and better performance. Figure 8.21 provides the performance analysis generated by Orange using the breast cancer dataset.

The model was constructed using fivefold cross-validation with 66% training dataset. The performance of each model is compared using the area under curve (AUC); cumulative accuracy (CA); F1 score, which is a weighted harmonic mean of precision and recall; precision; and recall. According to the results, NN has the highest CA, i.e., 71%, recall (77.2%), and precision (75%) with F1 score of 75%. F1 score is good when it is near to 100%. It refers model performance in combination of precision and recall. Overall NN is constructing a better model.

Figures 8.22a, 8.22b and 8.22c shows the confusion matrices of all the three algorithms. The dataset size was 290, out of which 55 cases belong to class 0 (No, not infected), and 235 cases belong to class 1 (Yes, infected).

Figure 8.22a shows the following results of SVM model:

- There are 4 cases which actually belong to class No and predicted also No.
- There are 51 cases which actually belong to class No but predicted as Yes.
- There are 9 cases which actually belong to class Yes but predicted as No.

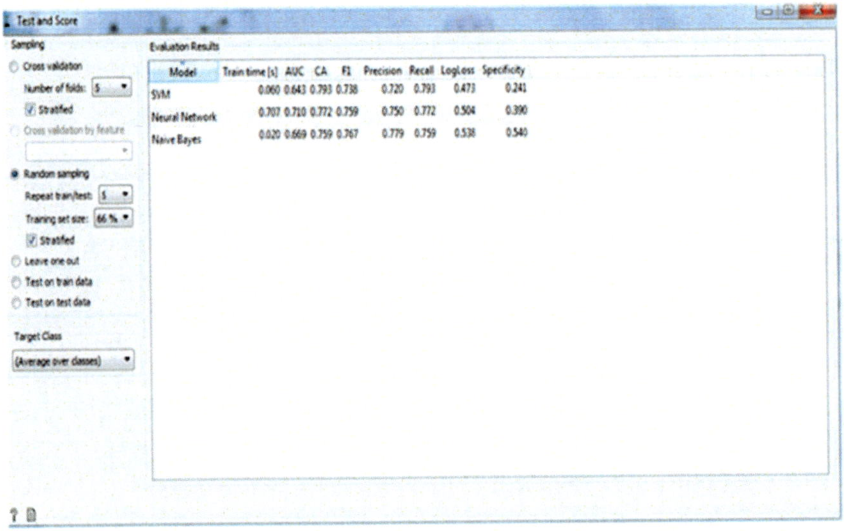

Fig. 8.21 Performance comparison of machine learning models for breast cancer dataset

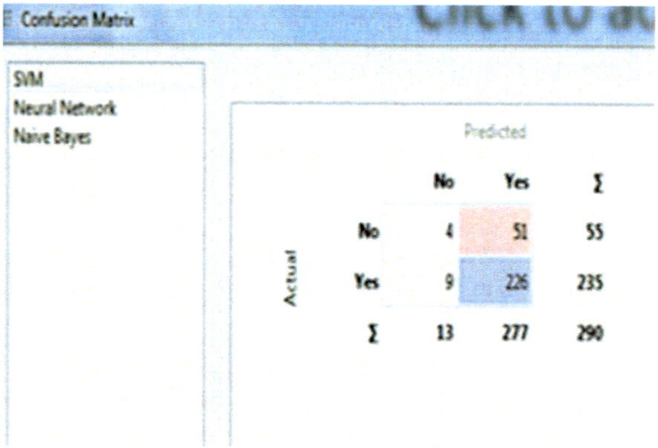

Fig. 8.22a Confusion matrix for breast cancer dataset using SVM

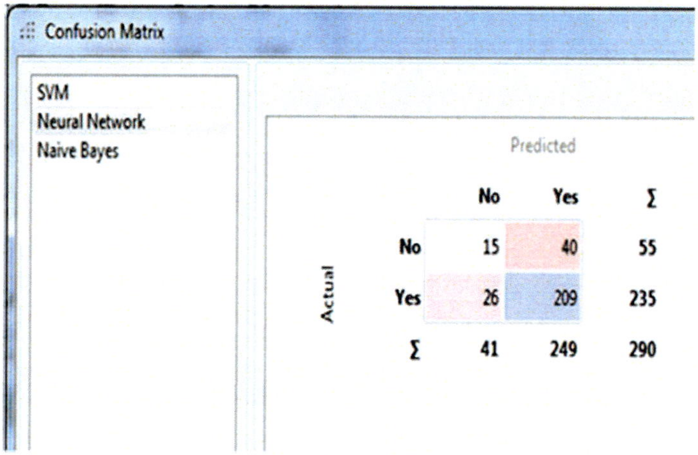

Fig. 8.22b Confusion matrix for breast cancer dataset using NN

- There are 226 cases which actually belong to class Yes and correctly predicted as Yes.

Therefore, out of 55 non-patients, SVM is able to predict only four cases, and out of 235 patients, SVM is able to predict only 226 cases accurately.

Figure 8.22b shows NN confusion matrix for the breast cancer dataset. The results are as follows:

- There are 15 cases which actually belong to class No and predicted also No.
- There are 40 cases which actually belong to class No but predicted as Yes.
- There are 26 cases which actually belong to class Yes but predicted as No.

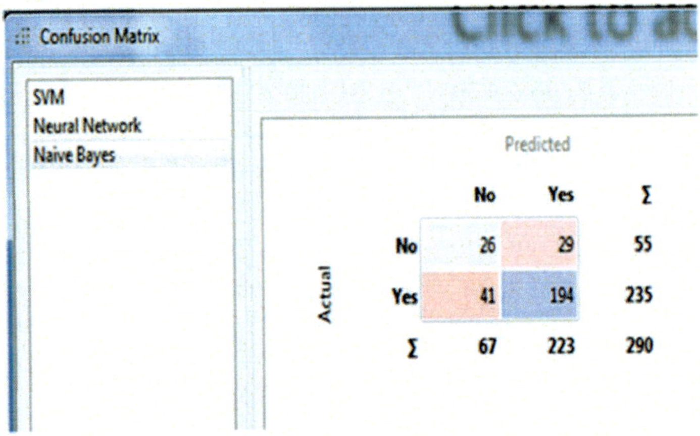

Fig. 8.22c Confusion matrix for breast cancer dataset using Naive Bayes

- There are 209 cases which actually belong to class Yes and correctly predicted as Yes.

Therefore, out of 235 patients, NN is able to predict only 209 cases, and out of 55 non-patients, NN is able to predict only 15 cases accurately.

Figure 8.22c shows the following results:

- There are 26 cases which actually belong to class No and predicted also No.
- There are 29 cases which actually belong to class No but predicted as Yes.
- There are 41 cases which actually belong to class Yes but predicted as No.
- There are 194 cases which actually belong to class Yes and correctly predicted as Yes.

Therefore, out of 235 patients, naive Bayes are able to predict only 194 cases, and out of 55 non-patients, naive Bayes is able to predict only 26 cases accurately.

These matrices reveal that NN and SVM are both appropriate models for breast cancer prediction.

Figure 8.23 shows the ROC curve, and the trade-off between the true positive and false positive rates are summarized for a predictive model.

- AUC of SVM is 0.500.
- AUC of Naive Bayes is 0.507.
- AUC of NN is 0.512.

The average value of the area under the curve (AUC) was observed in all the classification models, that is, neural networks (0.512), SVM (0.50), and Naive Bayes (0.507). Still, neural networks (NN) is a good prediction model for the breast cancer dataset.

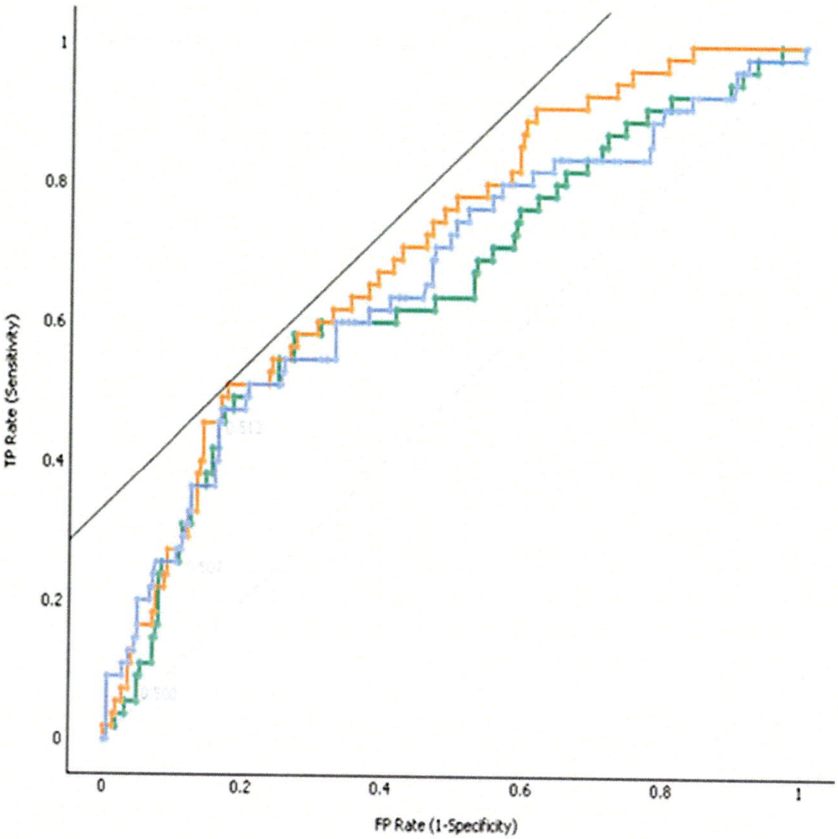

Fig. 8.23 ROC curve for breast cancer dataset

Further, using the RStudio, NN model was designed to show the three layers of neurons as shown in Fig. 8.24. This time, to check the model precision and accuracy, we take a random sample of 168 data from 290 dataset.

In Fig. 8.25, NN performance has been evaluated using 50–50 split and 70–30 split. This time the size of the test data in the 50–50 split was 84 (as the complete dataset was 168). In 84, four were of class No (shown as 0 in Fig. 8.25) and 58 were of class Yes (shown as 1 in Fig. 8.25). Briefly, the NN performed at 96% accuracy.

Similarly in the case of 70–30 split, the size of the test dataset was 52, where two were of class No and 39 was of class Yes. NN accuracy was 95%.

In short, the following observations were registered:

- Neural networks outperforms in all the three datasets. Its accuracy was high in comparison to SVM and Naive Bayes.
- SVM precision was better in all the results.

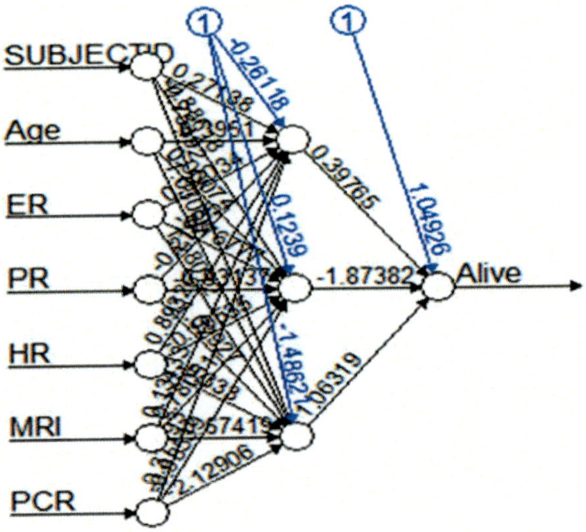

Fig. 8.24 NN model for breast cancer dataset using RStudio

- **For 50-50%, Train=1-84 and Test=85-168**

	prediction	
actual	**0**	**1**
0	**4**	**9**
1	**13**	**58**

- **For 70-30%, Train=1-116,117-168**

	prediction	
actual	**0**	**1**
0	**2**	**5**
1	**5**	**39**

Fig. 8.25 Classification matrix of neural networks model by RStudio

- Also the dataset suffers from unbalanced class problem. In each dataset patients or infected number was much higher than noninfected cases. The results will definitely change when balanced data is supplied to these models.
- Cross-validation is another important aspect in verifying the results. In this study, fivefold and tenfold validations are taken, from which the best results were selected for discussion.

- Orange tool has provided clear visualization of the curve and confusion matrices. It has helped to understand the performances of each model.

8.5 Major Findings and Issues

- There are several machine learning algorithms presented in order to analyze different types of cancer datasets.
- The main aim in the machine learning (ML) field was to construct precise classifiers for medical dataset usage.
- In this study, three algorithms have been used such as SVM, NN, and Naive Bayes on different types of cancer.
- These algorithms have been compared in order to find the best classifiers in terms of accuracy, specificity, and time taken to construct the model.
- Hence, the neural networks classifier has reached the highest accuracy and excelled all other classifiers.

8.6 Future Possibilities and Challenges in Cancer Prognosis

The present study has the future potential to apply ML models in other data with different features, related to survival prognosis of the patients. Machine learning algorithms have become a significant technique for a variety of applications in astronomy, social media, medical diagnostics, online trading, smart devices, online education, etc. (Mitchell 1997; Duda et al. 2001). The ML algorithms are powerful from the traditional problem-solving algorithms with the ability to learn from the data without being explicitly programmed. With the advent of cloud computing, the data management and storage issues can be handled with greater ease and flexibility, while the data analytics part can be well addressed by the use of machine learning algorithms. Medical science has exponential dataset and finds ML very useful for early diagnostic or prescriptive analysis (Islam et al. 2020). The future possibility of ML in cancer prognosis is:

- Precision medicine.
- Gene-based analysis for cancer generation.
- Emotional aspect of human and cancer susceptibility.
- Drug-target interaction and identification of natural drugs for cancer.
- Recommendation systems for symptom analysis, disease detection, and treatment prescription.

Challenges in devising these systems are high. ML itself is not sufficient enough to lead all of these systems. Merging of Internet of things (IoT) technology with ML is necessary. But IoT-based systems are complex and costly. Another crucial challenge with the convergence of AI in cancer prognosis is the privacy and data security issues. The prevailing problems of data breaching and hacking make the use of ML algorithms less preferable as the details of personal medical history of the patients are at risk of leaking. Moreover, the deliberate hacking of the algorithms can harm the patients at a large scale (Topol 2019). The algorithms are even susceptible to the risk of adversarial attack or manipulation by the inputs that are explicitly designed to fool them (Finlayson et al. 2019). Also, to handle data security issues, block chain has become very popular. It has also been introduced into the supply chain of pharmaceuticals. In order to induce data security in the online recommendation system, merging of block chain with ML is the need of the hour. The application of ML algorithms will result in a paradigm shift in cancer diagnosis and prognosis as the survival rates of the patients will be dramatically improved. The foreseeable future will include numerous advances in the ML algorithms that will resolve the current challenges.

References

Ahuja AS (2019) The impact of artificial intelligence in medicine on the future role of the physician. PeerJ 7:e7702. https://doi.org/10.7717/peerj.7702

Baralt LB, McCormick S (2010) A review of advocate–scientist collaboration in federally funded environmental breast cancer research centers. Environ Health Perspect 118:1668–1675. https://doi.org/10.1289/ehp.0901603

Barlow H, Mao S, Khushi M (2019) Predicting high-risk prostate cancer using machine learning methods. Data 4:129. https://doi.org/10.3390/data4030129

Crissien AM, Frenette C (2014) Current management of hepatocellular carcinoma. Gastroenterol Hepatol (N Y) 10:153–161

Cruz JA, Wishart DS (2007) Applications of machine learning in cancer prediction and prognosis. Cancer Inform 2:59–77

Delen D, Walker G, Kadam A (2005) Predicting breast cancer survivability: a comparison of three data mining methods. Artif Intell Med 34:113–127. https://doi.org/10.1016/j.artmed.2004.07.002

Duda RO, Hart PE, Stork DG et al (2001) Pattern classification, 2nd edn. Wiley, Hoboken, NJ

Ferlay J, Shin H-R, Bray F et al (2010) Estimates of worldwide burden of cancer in 2008: GLOBOCAN 2008. Int J Cancer 127:2893–2917. https://doi.org/10.1002/ijc.25516

Finlayson SG, Bowers JD, Ito J et al (2019) Adversarial attacks on medical machine learning. Science 363:1287–1289. https://doi.org/10.1126/science.aaw4399

Fortunato M, Azar MG, Piot B et al (2019) Noisy networks for exploration. arXiv:170610295 [cs, stat]

Hanahan D, Weinberg RA (2011) Hallmarks of cancer: the next generation. Cell 144:646–674. https://doi.org/10.1016/j.cell.2011.02.013

Islam MM, Haque MR, Iqbal H et al (2020) Breast cancer prediction: a comparative study using machine learning techniques. SN Comput Sci 1:290. https://doi.org/10.1007/s42979-020-00305-w

Kourou K, Exarchos TP, Exarchos KP et al (2015) Machine learning applications in cancer prognosis and prediction. Comput Struct Biotechnol J 13:8–17. https://doi.org/10.1016/j.csbj.2014.11.005

Lundin M, Lundin J, Burke HB et al (1999) Artificial neural networks applied to survival prediction in breast cancer. Oncology 57:281–286. https://doi.org/10.1159/000012061

Mitchell TM (1997) Machine learning. McGraw-Hill, New York

Mitchell TM (2006) Machine learning. McGraw-Hill, New York

Murali N, Kucukkaya A, Petukhova A et al (2020) Supervised machine learning in oncology: a clinician's guide. Dig Dis Interv 4:73–81. https://doi.org/10.1055/s-0040-1705097

Nagy M, Radakovich N, Nazha A (2020) Machine learning in oncology: what should clinicians know? JCO Clin Cancer Inform 4:799–810. https://doi.org/10.1200/CCI.20.00049

Obafemi O, Stephen A, Ajayi O, Nkosinathi M (2019) A survey of artificial neural network-based prediction models for thermal properties of biomass. Procedia Manuf 33:184–191. https://doi.org/10.1016/j.promfg.2019.04.103

Obaid OI, Mohammed MA, Ghani MKA et al (2018) Evaluating the performance of machine learning techniques in the classification of Wisconsin breast cancer. Int J Eng Technol 7:160–166. https://doi.org/10.14419/ijet.v7i4.36.23737

Pendharkar PC, Rodger JA, Yaverbaum GJ et al (1999) Association, statistical, mathematical and neural approaches for mining breast cancer patterns. Expert Syst Appl 17:223–232. https://doi.org/10.1016/S0957-4174(99)00036-6

Sayed S (2018) Machine learning is the future of cancer prediction. In: Medium. https://towardsdatascience.com/machine-learning-is-the-future-of-cancer-prediction-e4d28e7e6dfa. Accessed 6 Sep 2020

Topol EJ (2019) High-performance medicine: the convergence of human and artificial intelligence. Nat Med 25:44–56. https://doi.org/10.1038/s41591-018-0300-7

Zhu W, Xie L, Han J, Guo X (2020) The application of deep learning in cancer prognosis prediction. Cancers (Basel) 12:603. https://doi.org/10.3390/cancers12030603

Use of Artificial Intelligence in Research and Clinical Decision Making for Combating Mycobacterial Diseases

9

Abstract

Tuberculosis (TB) and leprosy (caused by mycobacterial pathogens) are two age-old infections, which we are facing even today. India is a major contributor to the global burden of leprosy and tuberculosis, which adversely affects the diverse communities as well as having a prevalence in different parts of the country. Timely diagnostics and effective treatment are very challenging, and the emergence of drug resistance has further complicated the management of these mycobacterial diseases. Various lineages of these mycobacterial pathogens show varying phenotypes in terms of clinical presentations and treatment outcomes. Altogether these factors make it further difficult to understand the full genetic diversity and pathogenicity of these pathogens using the conventional genomic and proteomic approaches. However, thanks to the recent technological advances in the genomics and proteomics field, many of these constraints have been suitably addressed. While it is relatively simpler to produce the omics data in a high-throughput manner, the bottleneck now is the pace to assimilate this large data into some useful information to reach a relevant, meaningful conclusion in a timely manner to assist the clinician in making a judgment.

In India, genetic diversity of different strains has been widely studied using approaches based on Next-generation sequencing (NGS), metagenomics, spoligotyping, and PCR. But there are still gaps in predicting phenotypes accurately from genotypic data, in particular for certain drugs. Recently, Machine learning (ML) methods were successfully used to develop predictive classification models and to identify compounds based on their biological activities. Artificial Intelligence- (AI) based ML learns from known data characteristics and makes predictions. Machine learning approaches can find statistical dependencies in the data and also take into account the non-linear and feature-interaction effects. In this way, new knowledge can be unleashed and data has been proven to be useful that can provide clinically actionable recommendations

A. Saxena, S. Chandra, *Artificial Intelligence and Machine Learning in Healthcare*,
https://doi.org/10.1007/978-981-16-0811-7_9

and high priority features like mutation/variant/polymorphism profile and its association with the drug as well as drug resistance profile, genotype information regarding clustering and molecular epidemiology of mycobacteria. Moreover, the data utilized by the model for prediction can also be implied in rapid diagnostics and transmission dynamics studies. In this chapter, we gathered the current information about the use of Genome-wide Association Study (GWAS) and NGS in mycobacterial disease and a machine learning literature supporting applications for identification and antimicrobial susceptibility testing in mycobacteria. We have attempted to provide a comprehensive introduction about the technological advancements in high throughput data and explain how NGS with ML can be used in clinical decision-making, genomics, proteomics, docking, simulations, drug screening, and drug-repurposing.

Keywords

Next generation sequencing (NGS) · Machine learning (ML) · Artificial intelligence (AI) · Genome-wide association study (GWAS) · Tuberculosis (TB) · Leprosy

9.1 Introduction of Technological Advancements and High Throughput Data in Genomics and Proteomics Work

Over the past sesquidecade, the development of high-performance molecular technologies and related bioinformatics has changed scientific capabilities dramatically in the processing, handling, and evaluation of large quantities of genomic, transcriptomic, and proteomic data (Manzoni et al. 2018). Techniques, such as the high-performance sequencing for gene and protein profiling, have transformed biological science into systematic surveillance of a bio-system. Regardless of the process, the high-performance data processing typically provides a list of genes or proteins expressed differently (Xia et al. 2014). This list is especially useful in the identification of the genes with functions in a specific condition or phenotype. Moreover, such approaches have been frequently used to classify complex biochemical structures, to investigate pathophysiological processes as well as identifying specific biomarkers (Pedlar et al. 2019; Lv et al. 2020; Dagiasis et al. 2014). The amount of DNA sequence data now available using NGS platforms is a clear example of this step change (Wadapurkar and Vyas 2018). Such biotech developments are increasingly being used for research in mycobacterial diseases and have started to revolutionize the molecular way in which biological and evolutionary processes can be studied. Next-generation platforms offer an unparalleled performance that produces giga-bases of data in a single experiment (Gupta and Gupta 2014; Thakur and Varshney 2010). Besides, these technologies provide the unbiased sequencing of the complete DNA, RNA, or protein content in a sample without prior knowledge and with the versatility to enable targeted sequencing and allow the detailed analysis of host-pathogen interactions at the level of their genomes

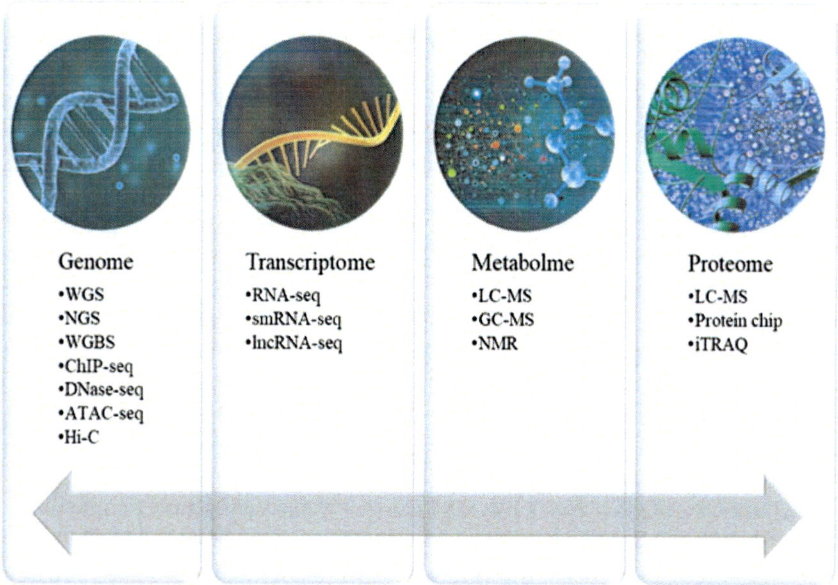

Fig. 9.1 The picture displays the interconnected gene expression domains, from genome to metabolite. Using microarrays, sequencing, and Mass spectrometry at each stage reveals to get multi-level gene and protein expression, these techniques delivered a multidimensional view of both natural and pathological processes

(genomics), transcriptomes (transcriptomics), proteomes (proteomics), and metabolome (metabolomics), respectively (Wanichthanarak et al. 2015) (Fig. 9.1).

9.1.1 High Throughput Screening of Tuberculosis

WHO Global Report 2018 claims that 27% of new cases of tuberculosis (TB) are from India, which is the highest number among countries with high TB burdens, followed by China with 9% of new cases (World Health Organization 2018). While the number of new cases worldwide was lower than those reported in the 2017 study, only a small shift was observed in the number of new cases coming from India. India also has the second highest incidence of cases of multidrug resistance (MDR), with the highest mortality rate (Lohiya et al. 2020). While the drug-susceptible cases had a higher rate of cure, the treatment success rate for MDR and XDR TB cases was just 54% and 30%, respectively. Moreover, the 2016 WHO global TB report estimated the prevalence of Rifampicin resistant TB (RR-TB) cases to be 4.1% in the new cases and 19% in the previously treated cases. Despite an increase in the number of testing for Rifampicin resistance, an estimated 240,000 people died from RR-TB (Chatterjee et al. 2018). In India, too, the number of cases of rifampicin resistant TB (RR-TB) is alarmingly high (Singh et al. 2020). Given the high cases of RR-TB in

India, there are only three major studies to date that focused on MTB's genetic diversity using a WGS-based approach to whole genome sequencing (Manson et al. 2017; Chatterjee et al. 2017; Advani et al. 2019). The majority of clinical isolates were drug-sensitive in the first two studies that restricted the detection of resistance mutations. But both studies emphasize the need for a diagnostic approach based on next-generation sequencing (Manson et al. 2017; Chatterjee et al. 2017). The third study conducted by ICMR-JALMA focused on MDR TB samples and discovered over 300 SNPs in 38 genes associated with drug resistance that are not used in diagnostic research (Advani et al. 2019). The study also found that bedaquiline-resistant mutations were present in seven MDR samples, including three from the Manson dataset. Other than these studies, there are several separate reports from clinical isolates with 2–3 TB WGS representing special cases, such as severe drug resistance (Rufai and Singh 2019; Kalo et al. 2015). However, to date, there are no such programs from India to compile all available data sets and provide the Indian environment with actionable therapeutic and diagnostic insights.

However, worldwide, numerous studies have comprehensively analyzed the perturbations in many tuberculosis virulent strains at the transcriptome and proteome level, such as identification of non-coding and micro RNAs, gene expression profiling of reference and mutant strains, and transcriptional start site mapping of clinical isolates (Tagliani et al. 2021; Wan et al. 2020; Romanowski et al. 2020). High throughput transcriptomic techniques such as microarrays and RNA sequence (RNA-seq) can assess the transcriptional response to the changes in mycobacterial genomes, such as nutrient starvation, antibiotic exposure, insufficient oxygen, etc. (Peng et al. 2020; Kwan et al. 2020; Hu et al. 2020; Liu et al. 2020a).

In a longitudinal study, a predictive signature for active TB disease had recently been applied with the objective of transcriptomic profiling. For two years, Zak and his colleagues have tracked healthy teenagers from South Africa, taking blood samples every six months (Zak et al. 2016). Forty-six people were finally diagnosed with TB from a total of thousand participants in the study. Transcriptomic profiles were collected from the blood samples of these individuals and compared with profiles of the individuals who remained healthy during this study prior to their time of diagnosis of TB and they were able to differentiate between the two groups with the Statistical significance.

In several studies, microRNA profile variations between TB patients and healthy controls, either from RNA derived from peripheral blood cells or from freely circulating microRNA present in patient serum or plasma samples, were examined (Wu et al. 2012; Chakrabarty et al. 2019; Spinelli et al. 2013; Yi et al. 2012; Qi et al. 2012). Nevertheless, it is still difficult to accurately interpret clearly lists of identifiers from biological features, because of our unfinished awareness of microRNA functions. Computational and experimental evaluation of biomarkers candidates shows that micro RNAs may play an important role in controlling immune response, for example, by affecting neutrophil mobilization in the lung (Dorhoi et al. 2013).

Proteomic analysis generally, has been performed with liquid-chromatographic tandem mass spectrometry (LC-MS) of cellular and secreted fractions, accompanied by study of uniform spectral counting such as by measuring normalized spectral abundance factor (NSAF) provides an improved measure for relative abundance, by factoring the length of the protein into subsequent calculations (Mehaffy et al. 2018). For example, in 2018, Mehaffy et al. used two separate MTB clonal pairs representing a particular genetic lineage (one clinical and one developed in the laboratory) but sharing a katG mutation related to INH resistance. Overall, after gaining INH resistance in both MTB genetic lineages studied, they have found 26 MTB proteins with altered abundances. These proteins were known to participate in the processes of virulence, lipid metabolism, detoxification, ATP synthesis, and adaptation (Mehaffy et al. 2018). Recently, the lack of proteomic data for various MTB H37RA genes has been reported in the study, with some attributed to virulence and pathogenicity mechanisms. Transcriptional and proteomic evidence for 3900 genes representing 80% of the estimated total gene count, including 408 non-identified proteins were found. Nine genes with no coding potential in H37Ra were also found, which include two supposed ESAT-6 virulence factors. In addition, proteogenomic analysis allowed 63 new gene-coding proteins to be identified (Pinto et al. 2018).

The effects of antibiotics on *M. tuberculosis* physiology have been supported by antibiotic improvements in gene expression profiles. Collectively, genes or group of genes fostering antibiotic resistance are called resistome. Variations in profiles of gene expression caused by antibiotics have enabled us to understand the impact of antibiotics on *M. tuberculosis* physiology (McNerney et al. 2018; Joshi et al. 2013). The drug-induced gene expression profile can be regarded as a transcriptional hallmark feature of the mode of action. These hallmarks can be used to predict the activity and mode of action of the novel/new anti-mycobacterial compounds. However, in order to predict the modes of action of new drugs based on comparisons with the expression profile of well defined compounds, the quality of the expression data is crucial.

In total, depending on the question asked, these high throughput technologies can be used in different ways. It can be used to examine changes in bacteria's gene-expression profile following the exposure to antibiotics in comparison to untreated cells, mutants' gene-expression profile in comparison to wild type cells treated with antibiotics, or clinical strain transcription profile, particularly in DR, MDR, or XDR strains. The Genome-wide profiles facilitate the characterization of action mechanisms and antimicrobial resistance mechanisms of the mycobacteria.

9.1.2 High Throughput Screening of Leprosy

Leprosy is caused by an uncultivated pathogen, *Mycobacterium leprae and Mycobacterium lepromatosis*, which primarily affects skin, mucosal surface of upper respiratory tract and the peripheral nerves (Bhandari et al. 2020). Nearly 250,000 new leprosy cases were reported from 131 countries, with 95% of those detected mainly in India, Brazil, Indonesia, and 20 other global priority countries (WHO, 2019). With over 1.25 lakh new leprosy cases detected in 2019, India accounts for

>60% of the total cases reported globally indicating an active transmission (Rao and Suneetha 2018). Leprosy diagnosis is mostly based on clinical presentations, and there is a great need of a suitable, field-friendly laboratory tool for assisting in its early and differential diagnosis. Repetitive loci (called RLEP, is present in 37 copies in *M. leprae* genome) is a preferred target for specific and sensitive detection of *M. leprae* DNA in clinical samples (Cole et al. 2001). In addition, appropriate tools for molecular epidemiology of leprosy are lacking. *M. leprae* strains from around the world have been classified on the basis of four SNP types (branches 1–4) and 16 SNP subtypes (1A–1D, 2E–2H, 3I–3 M, and 4 N–4P) based on comparative genomic analysis of four different *M. leprae* strains from India, Brazil, Thailand, and the United States which require PCR-sequencing of several genomic loci, making it very challenging due to limited amount of genomic DNA of the pathogen from the clinical samples (Monot et al. 2009). Genotyping a large panel of *M. leprae* strains has revealed its strong geographical association, thereby suggesting possible routes of dissemination worldwide. However, there is a very limited genomic information currently available about *M. leprae* strains present in India (Benjak et al. 2018).

Previous high throughput SNP typing studies of *M. leprae* from various endemic regions in India have shown that the SNP subtype 1D is the most prevalent genotype in India (present in ~76% of the cases), while other SNP types are 1B, 1C, 2E, 2H, and 2G (Monot et al. 2009; Lavania et al. 2015). The emergence of multidrug-resistant in *M. leprae* is also a major concern (Lavania et al. 2018; Matsuoka 2010). The molecular epidemiology of leprosy is challenging as it requires PCR-Sequencing of multiple loci (Scollard et al. 2006).

Whole Genome Sequencing was also recognized as an effective genotyping method, as it allows for a finer resolution of the genetic diversity of each isolate and offers the best dataset for population-based research (Monot et al. 2009; Lavania et al. 2015). In 2009, there were only four complete genomes of leprosy, but this small quantity of strains led to a good typing method and astounding data on strain variations and genetic evolution. Since 2009, along with 16 subtypes, new subtypes have increasingly been reported, for example, a study reported the new genotype called 1B-Bangladesh (Tio-Coma et al. 2020).

In 2011, *M. leprae* was reported to be entirely re-sequenced from a wild armadillo and three patients with leprosy in the US. Comparative genomic analysis between Asian and Brazilian strains revealed 51 SNPs and 11-bp insertion-deletion. The *M. leprae* genotype of foreign exposure patients usually represented their country of origin or history of travel. In 28 out of the 33 wild armadillos and 25 out of the 39 US patients who were living in areas of armadillo-borne *M. leprae*, a single and previously not reported M. *leprae* genotype (3I-2-v1) was found (Truman et al. 2011).

Similarly, in 2013, the *M. leprae* genome was sequenced from five Medieval skeletons from UK, Sweden, and Denmark using the DNA array capture (Schuenemann et al. 2013). The old *M. leprae* sequences were compared with 11 contemporary strains of different genotypes and geographical origins. Comparisons revealed that over the past thousand years the conservation was remarkable, that leprosy is European in the Americas, and that the *M. Leprae* genotype in medieval Europe is common with the Middle East, which has produced a significant impact on the study of palaeomicrobiology and evolution of human pathogens.

Consequently, in 2015, a thorough evaluation was made with the use of micro-arrays of DNA chip, covering the entire spectrum of the disease together with its reactional states, of human mRNA for leprosy skin lesions. Sixty-six leprotic (10TT, 10BT, 10BB, 10BL, 5LL, 14R1 and 10R2) samples and nine safe skin biopsies containing healthy males and females were used as controls. In this study, 1580 mRNA were found to be differentially expressed in diseased lesions versus healthy controls. Also, several genes have been found in all leprotic cases, whereas other genes were found in reactional states only, such as Type "1": GPNMB, IL1B, MICAL2, FOXQ1; type "2," AKR1B10, FAM180B, FOXQ1, NNMT, NR1D1, PTX3, TNFRSF25 (Belone et al. 2015). The role of these mRNAs have been explored in developing new diagnostic markers and therapeutic targets for leprosy as these mRNAs are known to be involved in various pathophysiological and signaling processes and in several other diseases (Mehta and Liu 2014).

Another important study in 2015, using deep sequencing, illustrates that the genomic sequence of *M. lepromatosis* present in a skin biopsy was linked with *M. leprae* that has undergone an extensive reduction. The genomes show broad synthesis and close in size (~ 3.27 Mb). Protein coding genes share the identity of 93% nucleotide sequence, and pseudogenes were 82% the same. Phylogenetic comparisons and the Bayesian dating analysis suggested that the two leprosy bacilli are remarkably preserved despite their ancient separations and still have similar pathologies (Singh et al. 2015).

With increased high throughput screening in the field of tuberculosis, it was demonstrated that strain variations modulate virulence, immune phenotypes, and play a crucial role in antibiotic susceptibilities with differential drug resistance and adaptation. The advancement in molecular leprosy research with the advancement of genome sequencing types has strengthened and established a similar pattern. Recently, some hypermutated genes were identified by the comparative genomics of the 150 leprae genomes of different geographical areas and presumed to play a role in the drug resistance, pathogenesis, or host adaptation of the bacterium (Benjak et al. 2018). Although mutations in a resistant rpoB, folP1 and gyrA area were present as a characteristic hallmark for drug resistance, authors identified three highly muted genes (ribD, fadD9 and nth) in drug vs. susceptible strains that indicate their direct involvement in medication resistance or compensatory mechanisms. However, few genes have been strongly mutated, independent from the genotype of drug resistance, for example, ml0411, a serine-rich antigen belongs to the PPE family (Benjak et al. 2018). This summarizes that the problem of traditional typing systems for leprae could be easily addressed by an entire genome approach. The technological difficulties, price, and lengthy downstream analyses, however limited their use.

A variety of studies have been carried out over the years to describe the leprae proteome (Parkash and Singh 2012). A high-throughput proteomic approach was undertaken in 2008 that resulted in identification of nearly 250 new proteins for

M. leprae. One hundred and four proteins were detected in the cell wall, 98 proteins in the membrane fraction and 60 proteins were identified in the soluble/cytosol fraction (Marques et al. 2008). In a 2009 report, 1046 proteins were identified, including five proteins encoded with previously forecast pseudogenes, using Gel-LC-MS/MS, using a linear quadruple ion trap-Orbitrap mass spectrometer (de Souza et al. 2009). Metabolic profiles extracted from urine were calculated and it was found that the urinary metabolome could be used to distinguish endemic controls from untreated mycobacterial disease patients, as regulation in the urine of patients with RR before RR initiation was also different from RR-Diagnose.

Few literature studies on *M. leprae* mRNA expression are also reported. Bleharski and colleagues assessed the genes' expression of leprosy patients having polar forms of skin lesions (Bleharski et al. 2003). They found many up-regulated mRNAs linked to antigen processing as well as presentation in leprosy. A comprehensive assessment of leprosy lesions with microarrays was conducted for differentially expressed miRNAs. As the levels of RNA expression were modulated by MDT, the assessment of the RNA pattern of expression may be a good predictor for leprosy treatment. Of the 1605 *M. leprae* genes, 315 suggested twofold higher signal intensity, which includes the family of metabolic Acyl-CoA enzymes and medicinal metabolic enzymes possibly linked to *M. leprae* virulence. Diana et al. published a study that tells about the expression of pseudogenes in *M. leprae* and which were showing regulated expression in different conditions (Williams et al. 2009). A similar study has been conducted to identify the microRNAs of leprae and showed the regulation in different disease condition, like reactions, drug resistance, and according to RJ classification (Akama et al. 2009).

As *M. leprae* cannot be cultured, therefore scientists are facing the daunting task of assigning molecular and cellular roles to thousands of newly predicted gene products. With the advent of high throughput screening, now *M. leprae* reference genome has about 2699 annotated active genes, and at least 2041 proteins are predicted to be produced by it which were 1604 previously, with now lesser number of pseudogenes, i.e., 607, which was previously thought to be 1155. Despite considerable progress, the identification of many more promising proteins still needs to be performed. The investigators are looking forward to developing new methodologies for preventing nerve damage, effective leprosy treatment, and diagnosis of *M. leprae*.

9.1.3 High Throughput and Ultra-High Throughput Screening of Compound Libraries for Drug Discovery and Drug Repurposing

The discovery of drugs and medicines at the end of the twentieth century mostly focused on target methods (Zuniga et al. 2015). In order to identify potential drug targets, the identification of the mycobacterial whole genomic sequences and their strains has played a crucial role (Ioerger et al. 2013). Several compound groups have been identified by high-performance target-based screening. Some of them are still being established at the leading stage. For example, Targets include PanC, FtsZ, FadD32, gyrA, rpoB, folP LeuRS, InhA for *M. tuberculosis* and *M. leprae* (Chetty et al. 2017; Islam et al. 2017; Uddin et al. 2016; Waman et al. 2019). Once structural knowledge is available, virtual screening has become more common, 3D objectives can be used to test possible inhibitors (Gimeno et al. 2019). This approach provides the advantage of limited laboratory work and the opportunity to scan very large libraries of compounds. For example, *M. tuberculosis* drug target DprE1 (Zhang et al. 2018a). A large scale virtual screening was done and from around four million compounds, 41 compounds were classified as likely inhibitors (Wilsey et al. 2013). Six of the compounds were active against *M. smegmatis*, indicating that the method is useful. Recently, it was believed that it is necessary not only to concentrate on novel bioactive compounds but also to repurpose existing compounds to a new molecular target in an attempt to discover new inhibitors (Singh et al. 2019; Štular et al. 2016; Nagpal et al. 2020; Pushkaran et al. 2019; Rani et al. 2020). It would be significantly less intensive effort and enormous financial burden on traditional drug development procedures to repurpose a known bioactive compound, especially with its proven pharmacological properties (Pan et al. 2014). InhA is an isoniazid target and remains of interest to many groups (Štular et al. 2016; Pauli et al. 2013). The 3D pharmaceutical model was developed based on 36 InhA crystal structures, including wild InhA and drug-resistant mutants InhA, apo InhA, and complex InhA, with either NADH, substratum, or ligand. Parallel to the quest for ligands, four docking programs and almost one million compounds have been screened; 19 molecules have been identified as possible noncytotoxic inhibitors. The enzyme was tested with six molecules and three inhibiting InhA purified molecules, though data have not yet been documented against living bacteria (Pauli et al. 2013).

The use of drug screens against individual patient isolates is an alternate approach to finding successful therapeutics against multidrug-resistant bacterial infections. It takes around 10–12 years on average with sufficient resources for the creation of a new antibiotic (Jackson et al. 2018) (Fig. 9.2).

For example, promising antibiotics have been found against MDR *Mycobacterium tuberculosis, Acinetobacter baumannii, and Borrelia burgdorferi* by recycling existing medicines (Sun et al. 2016; Silva et al. 2018). There are also thousands of additional approved antibiotics for illnesses other than infections that can be administered against MDR bacteria or can potentially resensitize MDR bacteria to standard care antibiotics by overcoming a specific medical resistance mechanism. Reports have identified <200 approved antibiotics available to clinicians to choose treatments. Current antitubercular therapy suffers from a longer-term disadvantage that presents a significant challenge to the growth of patient non-compliance and resistance. The current situation needs alternative approaches, which can reduce care time so that improved health results can be achieved. For example, drug repurposing and medications, namely, statins, metformin, Bevacizumab, Zileuton, ibuprofen, aspirin, Valproic acid, Adalimumab, and Vitamin D3, have shown promising results in clinical outcomes in TB patients during preliminary examination (Mishra et al. 2020). The key benefit of this drug repurposing screening strategy is to recognize and apply licensed drugs with a new identity. Antimicrobial compounds may pass

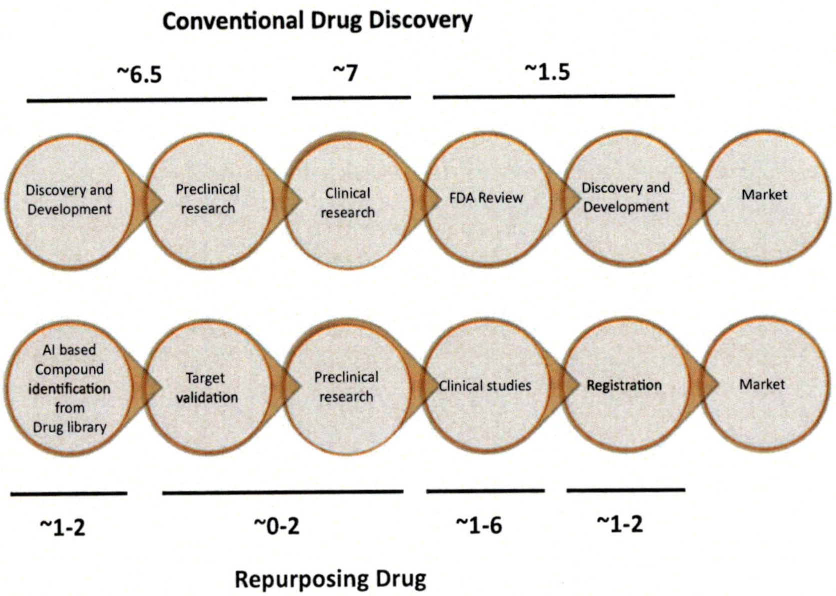

Fig. 9.2 Schematic representation of the steps involved in traditional drug discovery process vs. AI based drug repurposing with the salient features of both the processes

quickly through clinical trials or therapies without a lengthy period of preclinical drug creation. Primary screening and validation of active compounds may also be completed within 1–2 weeks (Sun et al. 2016). In this way, the existing drug can be used to treat other symptoms based on the target molecule.

9.2 High Volume Data and the Bottleneck in Data Analysis

9.2.1 Development of Omics Data

With the emergence of the genomic era, the use of high-throughput genomics have started to generate biological data at an exponential pace (Chance et al. 2004; Esfandyarpour et al. 2013; Lebrigand et al. 2020; Sarnaik et al. 2020). The scientific field of -omics provides vast volumes of data primarily on the basis of advances in genomics and biotechnology (Oliveira 2019; Jiang and He 2020). High-throughput systems that calculate the expression of thousands of genes or non-coding transcripts (e.g., miRNAs), genotyping methods and next-generation sequencing (NGS) technologies, whole genome-wide interaction studies (GWAS) that produce quantitative gene expression profiles (e.g., RNA-seq), identification of a significant number of gene variants (SNPs, Indels); are some of the major applications (Koumakis 2020; Zhang et al. 2017; Qin 2019). The vast volume of data creates unprecedented possibilities for research at the genomic or systemic level, which opens the door for new biological findings (Fig. 9.3).

However, this modern paradigm faces severe challenges, like data accuracy, which must be monitored on the scale of the genome because analysis of data sets polluted with erroneous data is likely to lead to erroneous conclusions. For example, manual curation has been shown that MTB TlyA was involved in ribosomal biogenesis and the functional annotation were incorrect, not only in microbial and plant genomes but also in *M. tuberculosis* (Arenas et al. 2011). Similarly, in 2020, it was shown that in all the mycobacterial family, the protein annotated as HemN could not exhibit coproporphyrinogen III dehydrogenase (CPDH) activity and has been mis-annotated as HemN and therefore highlights the need to correct the present annotation to heme chaperone HemW in various bioinformatics databases (unpublished data). The main reason behind is a presence of a variety of protein sequence databases which appears to be polluted with incorrect/incomplete sequences. The reason behind lacking of proper scrutiny is the growing proportion of protein sequences derived from huge genome sequencing data, but since few genomes have been completely sequenced so far, researchers are annotating the sequences through comparative approaches, depending on sequence alignments (Prada and Boore 2019). However, in the case of genome design, sequencing errors, sequence gaps, and misassemblies result in an excessive rate of misannotations (Nobre et al. 2016; Wakeling et al. 2019). One significant cause of this error is that, in genomes, the apparent number of genes can be divided into several contigs that leads to the increase in the number of incorrect genes (Denton et al. 2014). Secondly, despite the completion of proper genome sequences and genome assemblies, the issue of protein coding genes prediction errors has emerged. In the case of intron-rich genomes, the ENCODE Genome Annotation Evaluation Project has shown clearly that the prediction of the correct structure of protein coding genes remains a difficult job (Guigó et al. 2006). Various approaches provided different predictions, but the most reliable were typically forecasting methods based on experimentally determined mRNA and protein sequences. Nevertheless, it was shown that the prediction of only about ~60% of the genes has an identical genomic structure of the protein-coding genes (Harrow et al. 2009). Most recently, a tool for exhaustive all-against-all sequence comparison called "Contaminator" has been described, which detected contamination in >2 million sequences (and 6795 species) in GenBank database, >114,000 sequences (in 2767 species) in the NCBI Reference Sequence Database (RefSeq), and 14,132 protein sequences the non-redundant (NR) protein database. These could be due to mislabeled/incorrectly labeled reference samples, contamination, or due to the presence of more than one species in some samples. As the sequence volume keeps on increasing, it is important to identify such sources which can cause false interpretations and resultant false interpretations (Steinegger and Salzberg 2020).

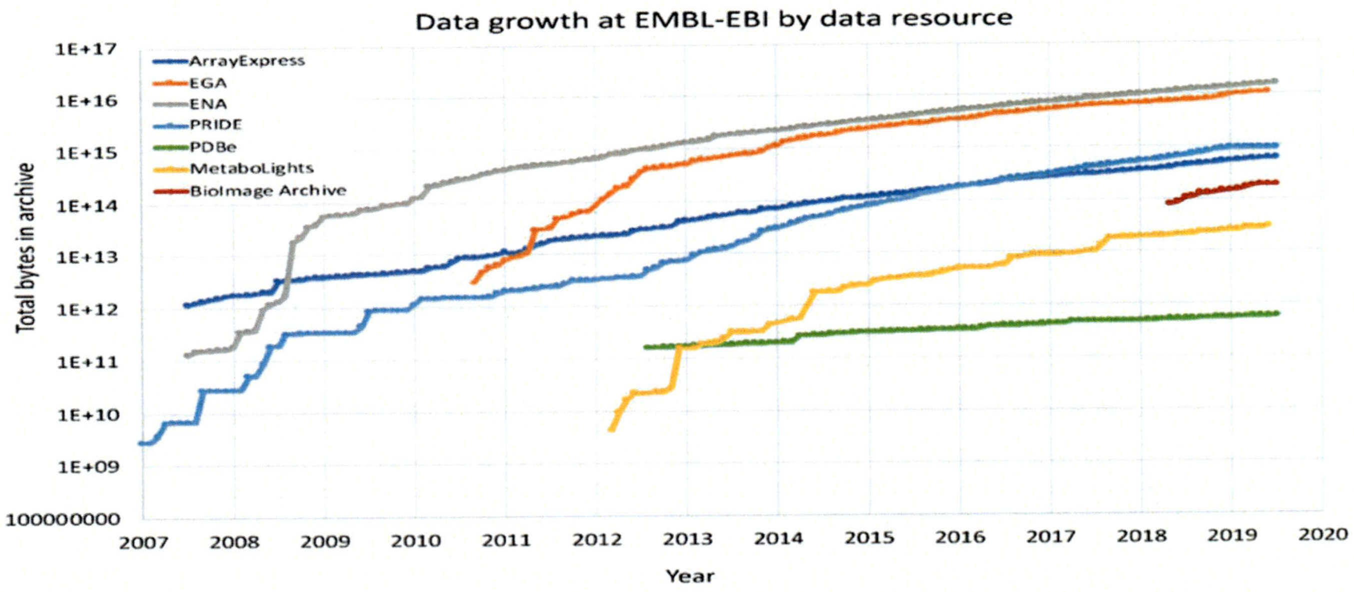

Fig. 9.3 Data accumulation at EMBL-EBI by data resource over time. The y-axis shows total bytes for a single copy of the data resource over time. Resources shown are the BioImage Archive, Proteomics IDEntifications (PRIDE), European Genome-Phenome Archive (EGA), ArrayExpress, European Nucleotide Archive (ENA), Protein Data Bank in Europe and MetaboLights. The y-axis for both charts is logarithmic, so not only are most data types growing, but the rate of growth is also increasing. For all data resources shown here, growth rates are predicted to continue increasing. From Cook et al., NAR, 2020

9.2.2 NGS and its Use in Clinical Decision-Making, Proteomics, Docking, Simulations, Drug Screening (Repurposing of Drugs)

One of the advantages of NGS is to analyze hundreds and thousands or even millions of goals simultaneously. The clinical NGS in mycobacterial investigations is not only a diagnostic program but it is also widely used in the identification of mutation targets for the treatment of certain tuberculosis and leprosy and the identification of a high risk population (Qin 2019). In recent years, various drugs have been created to target molecules and more will be available. This capability provides NGS tremendous potential for clinical application. For example, any tumor can have multiple mutations in cancer patient treatment, any disease can have a number of SNPs involved and a number of pathways involved in the progression of the disease (Di Resta et al. 2018). In these clinical environments, typical molecular tests require multiple tests for many mutations. For these multiple tests, a larger amount of tissue may be required. Those targets can be challenged in a single test using NGS technology (Mokrousov et al. 2016; Eloit 2014). Therefore, less tissue is needed and tested results are obtained from dozens and hundreds of DNA targets. The number of mutations in different diseases has increased in recent years in scientific research. For example, numerous mutations were found in *Mycobacterium tuberculosis* and *Mycobacterium leprae* that lead to drug resistance, loss of function, pseudogene formation, loss of protein-protein interaction, etc. (Singh et al. 2020; Chatterjee et al. 2017; Wan et al. 2020; Benjak et al. 2018; Matsuoka et al. 2007; Singh and Cole 2011) These results also indicate that diagnostic and follow-up molecular trials should be conducted for multiple mutations. The burden of mutation has become a significant parameter to be evaluated with the introduction of immunotherapy (Kim et al. 2020). Numerous mutations in a TB and leprosy sample need to be investigated again. Typical molecular research procedures for these needs are not useful (Grossman et al. 2013). For certain tasks of patient care, NGS technology is therefore appropriate. In the current medical practice, more details on mutation must also be derived from biopsy samples (Hodgson et al. 2012). Since biopsy samples are very small, traditional molecular tests are often not possible to meet such requirements. In order to meet these needs, NGS was developed. NGS technology can test several samples and multiple targets simultaneously by massive parallel sequencing. This, therefore, increases molecular test processing time (Yohe and Thyagarajan 2017). In personalized precision medicine, it has become clear that NGS technology is an important tool. It offers information for the classification of disease conditions, therapeutic selection, and prognostic assessment. The use of NGS in clinical settings, however, entails difficulties (Bacher et al. 2018). For example, several reports have been made using NGS technology to disclose profiles of drug resistance in MTB. In prior studies, only one or more MTB drugs, which were resistant and without susceptible strains, were usually used. Nevertheless, it is extremely doubtful that this condition will arise in clinical practice. Without prior information on the resistor status, clinicians need to use checks, which mean that they need details about the relationship or non-relation of the variant found in clinical specimens. In this context, the distribution of each gene and healthy polymorphisms not linked to the drug resistance should be considered when evaluating NGS results (Kumar and Abubakar 2015).

9.3 Advent of Artificial Intelligence (AI) & Machine Learning (ML)

9.3.1 Machine Learning and Deep Learning (DL) Algorithms

Researchers are able to generate and interpret a large deal of omics data with the advancement of biotechnology and the advent of high-performance sequencing. Because, a high number of High-throughput data, sometimes known as "big" data, is generated, most of the algorithms in bioinformatics are focused on master learning and, recently (Lyko et al. 2016), on deep learning to recognize trends, predict the course of treatment of disease, and model it. Machine learning advances have created unprecedented momentum in biomedical computer science and have led to new fields of biological information and computational biology research (Camacho et al. 2018). Machine learning is an artificial intelligence division that focuses on algorithms and strategies for learning by examples by gathering characteristics of interest depending on the underlying distribution of probabilities (Rajkomar et al. 2019). It has the same idea as the expert system; it can mimic a human expert's capabilities. It can make an automated decision based on the knowledgebase the domain expert has entered. Since human expertise is not always accessible or sufficient to meet the community's needs, diagnostic software using machine learning can be used as a replacement for human expertise (Allam 2020).

It is evident that in specific tasks in omics data, machine learning models can have greater accuracy than state-of-the-art approaches (Lane et al. 2018). The increasing trend in deep learning architectures in genomic research, deep learning, and machine learning, particularly for multiscale and multimodal data analysis for precision media, is anticipating accelerated changes in genomics (Libbrecht and Noble 2015; Zou et al. 2019). Owing to huge data generation, the era known as "big" data, deep learning methods have shown to be an efficient discipline of ML. Machine learning techniques have successfully been used to develop predictive classification models, including compound recognition, based on their biological behaviors, predictions for side effects, new gene predictions associated with diseases, micro-array data processing, and drug development (Liu et al. 2013). AI-based ML learns from known data characteristics and then makes blind data predictions. In order to identify single nucleotide variants (SNV's) as immune or TB prone, Artificial Intelligence and ML algorithms have already been used to determine new mutation-supported resistance (Oliveira 2019).

There are various benefits of various ML algorithms. To that end, four algorithms have been predicted by supervised users, namely, naïve Bayes (NB), k next-door neighbor (kNN), artificial neural network (ANN), and sequential minimization (SMO) algorithm, based on Support Vector Machine (SVM) (Deepika and Seema

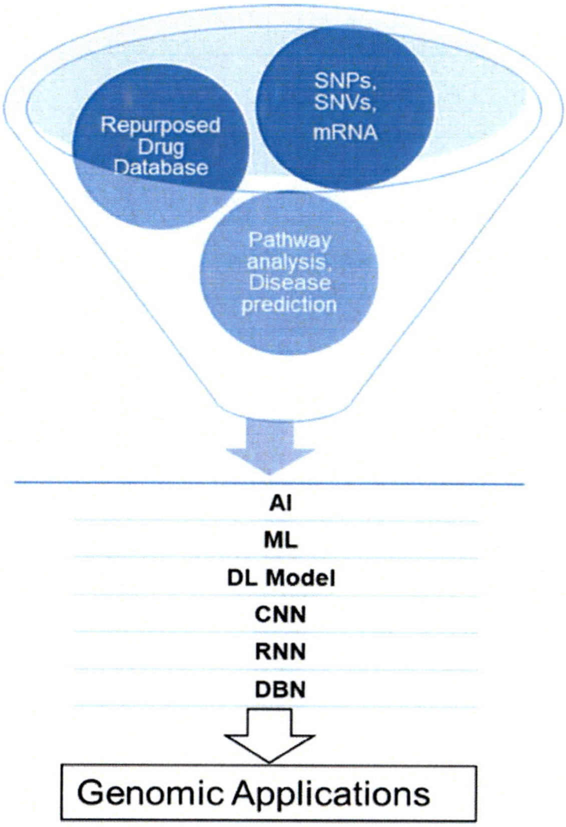

Fig. 9.4 Schematic representation of the steps involved in AI-based prediction models for genomic applications

2016). Deep learning algorithms include Convolutional Neural Network (CNN), Recurrent Neural Networks (RNN), Generative Adversarial Networks (GANs), Long short-term memory (LSTM), and Autoencoders (AE) (Munir et al. 2019). Methods may also be mixed to improve predictive performance with DL or ML models. The Multi-model Fusion is one such approach which includes meta-analysis of multiple models based on various data to achieve a common target. Decision fusion integrates the effects of several classifications into a single final forecast that forms a meta-estimator using statistical methods to amplify each classifier (Koumakis 2020). There are also sequential fusion models, including DanQ that use CNN, then RNN to calculate DNA sequence function (Zhang et al. 2019). Both contribute to increased predictive ability and may overcome inconsistencies or discrepancies in the specific analysis. These algorithms can be used to build prediction models (Fig. 9.4).

Further, the most accurate classification models in all tested genes can be assessed with an external invisible data set to reveal their applications. In addition, molecular docking and molecular dynamic simulations for wild type and forecast resistance can be performed, which will research the effect on protein conformation and trigger mutant protein and anti-TB drug complexes to validate the phenotype observed (Priya Doss et al. 2014).

9.3.2 AI in Drug Repurposing

The repurposing of already present drug substances for various indications can significantly reduce the time and cost needed to develop new medicinal products (Pushpakom et al. 2019; Oprea and Mestres 2012). While this field has graduated with a range of software tools from the discovery to the purposeful assessment, artificial intelligence progress is expected to dramatically improve predictive capability (Paranjpe et al. 2019). Taking advantage of the thousands of approved drugs and more than 4000 compounds abandoned during phase II production in new drug development activities is especially useful when aimed at neglected diseases like leprosy (Parvathaneni et al. 2019). Likewise, since many current antituber medications cause major side effects as well as promote resistance, it is very tempting to repurpose non-resistant agents with limited side effects into TB medicines (Passi et al. 2018). Advances in methods of drug repurposing and access to genomic data also allow the systematic development of personalized, repurposed options. Through machine learning models, computational drugs repurpose has moved to modern methods for analyzing drug effects using conventional biological approaches focused on determining chemical similarities and molecular dockings (Kinnings et al. 2011). Examples include gene expression and functional strategies focused on the genomics, such as corresponding drug indications by disease-specific response profiles on the basis of gene expression and mRNA expression. Another example includes identification of new possible protein target indications through genome-wide association studies (GWAS), generation of genetic variation-based approaches to find out Single nucleotide variations as a result of drug are some of the solutions provided by AI to find out the overall effect of drug in the system (Schneider 2018). These approaches are based on disease-networks that relate knowledge on diseases scrapped from different public resources to create multi-level networks (e.g., reactomes, KEGG text-mining pathways) or a disease graph based on gene expression profiles and protein networks. Due to the rapid accumulation and growing accessibility and standardization of chemical and genomic data alongside pharmacological and phenotypic knowledge, drug repurposing is becoming an excellent case study for proponents of the implementation of AI technologies in the pharmaceutical field (Mak and Pichika 2019). The question plays with AI's strengths in collecting insightful features from noisy, incomplete, and high-performance data. Different AI-based methods were suggested for identifying potential drug exploiting opportunities through the integration of diverse heterogeneous data sources information; examples include PREDICT, SLAMS, NetLapRLS, and DTINet (Yang et al. 2019). In field design, AI is implemented via the generation of the learning prediction model and performs a quick virtual screening to show the output accurately. Moreover, AI can easily identify drugs and can combat new diseases, including leprosy and tuberculosis, through a drug repurposing strategy.

This technology is indeed an evidence-based medical resource that can enhance the patient's identification, preparation, diagnosis, and is being research-based.

9.3.3 Examples from NGS and its Use in Clinical Decision-Making, Proteomics, Docking, Simulations, Drug Screening (Repurposing of Drugs)

One of the advantages of NGS is to analyze hundreds and thousands or even millions of goals simultaneously (Hodkinson and Grice 2015). The clinical NGS is not only a diagnostic program. It's also widely used in the identification of mutation targets for the treatment of certain tuberculosis and leprosy and the identification of a high risk population (Advani et al. 2019; McNerney et al. 2018; Monot et al. 2009). In recent years, various drugs have been created to target molecules and more will be available. This capability provides NGS tremendous potential for clinical application. Any tumor can have multiple mutations in cancer patient treatment, for example. In these clinical environments, typical molecular tests require multiple tests for many mutations. For these multiple tests, a larger amount of tissue may be required. Those targets can be challenged in a single test using NGS technology (Papadopoulou et al. 2019). Therefore, less tissue is needed and tested results are obtained from dozens and hundreds of DNA targets (Buyuksimsek et al. 2019). The number of mutations in different diseases has increased in recent years in scientific research. For example, numerous mutations were found in *Mycobacterium tuberculosis* and *Mycobacterium leprae* that lead to drug resistance, loss of function, pseudogene formation, loss of protein-protein interaction, etc. These results also indicate that diagnostic and follow-up molecular trials should be conducted for multiple mutations. NGS technology can test several samples and multiple targets simultaneously by massive parallel sequencing. This, therefore, increases molecular test processing time. In personalized precision medicine, it has become clear that NGS technology is an important tool. It offers information for the classification of disease conditions, therapeutic selection, and prognostic assessment. The use of NGS in clinical settings, however, entails difficulties. For example, several reports have been made using NGS technology to disclose profiles of drug resistance in MTB. In prior studies, only one or more MTB drugs, which were resistant and without susceptible strains, were usually used. Nevertheless, it is extremely doubtful that this condition will arise in clinical practice. Without prior information on the resistor status, clinicians need to use checks, which mean that they need details about the relationship or non-relation of variant found in clinical specimens. In this context, the distribution of each gene and healthy polymorphisms not linked to the drug resistance should be considered when evaluating NGS results.

9.4 Illustrations of Machine Learning in Different Research Fields

9.4.1 AI and ML in Covid-19-Related Research

The spread of COVID-19 produced a catastrophe, and the rapid treatment of this disease is a preventive medication with a history of patients recovered in the present pandemic (Fauci et al. 2020). In the COVID-19 scenario, the use of AI-enabled medication can be beneficial with technological advances in Artificial Intelligence (AI), together with increased computational resources (Vaishya et al. 2020). The pharmaceutical industry also seeks new and state-of-the-art technology in this respect to map, control, and limit the spread of COVID-19 disease (Swayamsiddha and Mohanty 2020; Ting et al. 2020). AI research models can be built to predict drug structures that can theoretically handle COVID-19 (Alimadadi et al. 2020). AI and machine learning can help the approach by quickly realizing that drugs have a sufficiency with COVID-19, and thus overcoming any barrier between a large number of drugs. A lot of information is available in open phases from various health services and organizations. A number of groups have started to use this advancement to increase the exposure of COVID-19 medicines and better understand the battle against infection by the resistant frame (Mohanty et al. 2020). GlaxoSmithKline (GSK) and Vir Biotechnology pharmaceutical companies joined forces to advance coronavirus treatment using computerized reasoning and CRISPR by early April. In addition, Harvard University was recently united with the Human Vaccines Project called Human Immunomics Initiative, which uses human-made thinking models to quicken antibodies to a wide range of infections, including COVID-19 (Mohanty et al. 2020). A knowledge representation system that uses GPS data to show users' locations of known COVID-19 cases has been lately developed by a team from Southern Illinois University (SIU). Google and Apple have worked to create a link with the Bluetooth software program (Mohanty et al. 2020). These methods can be very efficient and accurate in the collection of data. Organizations are carrying out research on various pathways in effectively accepted medicines, having identified human well-being profiles, based on a simple understanding of the infection (Shi et al. 2020). With regard to COVID-19, the two most popular instances of this are hydroxychloroquine (endorsed to treat malaria), remdesivir (Ebola). Therefore, an AI model can be modeled well by giving the input from the data set to find out the efficacy of these medicines (Mohanty et al. 2020). Likewise, groups of work started to look at artificial intelligence (AI) as a method to read and analyze XR and CT scans, and these forms of COVID-19 AI-based methods may be broadened to include all kinds of respiratory diseases. For example, Deep learning detects COVID-19 and separates it from pneumonia using chest CT, which means that AI could help turn a standard CT or X-ray scan into a versatile tool for prompt diagnosis, which would not only be useful for detecting COVID-19 but also other respiratory diseases (Li et al. 2020). In order to speed up potential COVID-19 case recognition, the use of ML algorithm via a mobile web-based survey was proposed that will reduce the dissemination of the

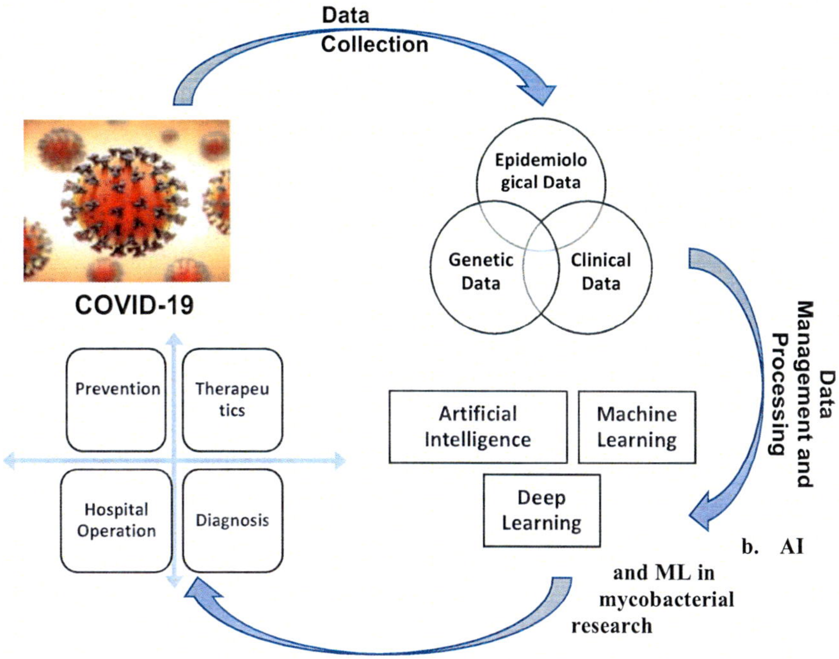

Fig. 9.5 The image depicts diverse applications of artificial intelligence in healthcare. The ability of AI to learn and rewrite its own rules, through Machine Learning and Deep Learning, offers not only benefits for today but also yet unseen capabilities for tomorrow

virus in vulnerable quarantine populations (Rao and Vazquez 2020). Israel's researchers have also developed the AI based Covid-19 test by using single sample of saliva with 95% accuracy rate that gives result in less than a second, known as Covid spit test (Israel21c 2020). In environments with limited diagnostic resources, such as rural or economically disadvantaged parts of the world, such quantification is particularly useful. In this regard, AI offers clear and actionable lung involvement details, providing an immediate risk assessment that is directly present on the X-ray (Mertz 2020). It is particularly useful to track the progression of the disease, to assess how well the patient responds to medication and to decide if improvements to medication might be appropriate. It may be safe to conclude that the Solutions from AI would help to make the average more expert (Fig. 9.5).

The emphasis is still on the development of new therapeutics for the fight against resistance to medications of first and second line of drugs used in TB treatment (Singh et al. 2020; Kalo et al. 2015; Kouchaki et al. 2019). It is of crucial importance to discover new TB-candidates with new mechanisms of action and shorter treatment duration. Much of the effort has been leveraged to large high-throughput screens in academia and industry, but the ratio of translating in vitro active compounds from these screens to in vivo is cumbersome as we have to find molecules that balance activity versus good physicochemical and pharmacokinetic properties (Prathipati et al. 2008). Work on the use of ML models for in vitro MTB datasets has contributed to the modeling of large MTB datasets, which have been made available for various classes (Lane et al. 2018). These models can be used to rate and filter similarly large numbers of molecules associated with pharmacophore methods before in vitro research. For example, in 2004, for media optimization, AI was used in the production of Rifamycin B via *Amycolatopsis mediterranei* S699 barbital insensitive mutant strain (Bapat and Wangikar 2004). Rifamycin B was considered to be an effective tuberculosis and leprosy antibiotic. To improve the medium composition, ML approaches were explored, such as genetic algorithm (GA), neighborhood analysis (NA), and decision tree technology. These medium combinations have increased Rifamycin B productivity by more than 600%, indicating that Genetic algorithms have become amazing at optimizing the fermenting medium and have qualitatively exposed the relationships between the media-media interaction in the form of collection of high, medium, and low productivity levels (Bapat and Wangikar 2004). Similarly, Bayesian models were used to predict several anti-tuber compounds. In 2014, by filtering the library of over 150,000 compounds, Bayesian models picked 48 compounds that can be tested in vitro; 11 were working with MIC values ranging from 0.4μM to 10.2μM, with high hit rate. These include five quinolones, three molecules with long aliphatic bonds and three singletons and, among these, were ciprofloxacin, a drug used to treat leprosy and tuberculosis (Ekins et al. 2014). A second validation of this method tested 550 molecules and 124 molecules were found active. A third example tested 48 compounds with an independent group and 11 were labeled as successful. A validation used a range of 1924 molecules as a comparison with the various ML models to demonstrate the enrichment rates which were in some cases greater than tenfold. Several experiments often analyze how MTB data sets are integrated and models of data reported by different groups are evaluated. For example, in 2018, a convolutional neural network-(CNN) based model was created to explicitly recognize the TB bacillus called TB-AI. Two hundred and one samples (108 positive cases and 93 negative cases) were gathered as the test set following the training of the neural network model to investigate TB-AI. TB-AI obtained a sensitivity of 97.94% and specificity of 83.65% against double confirmed diagnosis both by microscopes and digital slides by pathologists (Xiong et al. 2018). These combined efforts demonstrated the significance of several MTB models and also indicated important molecular characteristics for the active agents that recently reported the development of new antibacterial β-lactam with MTB activity. ThyX and Topoisomerase I have further established machine learning models for individual drug discovery targets (Djaout et al. 2016). In order to precisely diagnose and predict new cases of leprosy, Brazilian scientists recently have developed combined molecular and serological methods research using AI based random forest (RF) algorithms. All the asymptomatic SSS samples were obtained for 16SrRNA qPCR and the ELISA tests for LID-1 and ND-O-LID antigens. Statistical analysis showed anti-LID-1 sensitivity (63.2%), ND-O-LID (57.9%), qPCR SSS (36.8%) and microscopic diffraction (30.2%). But the use of RF suggests a strong increase in the sensitivity of MB leprosy (90.5%), PB leprosy (70.6%) with a 92.5% specificity (Gama et al. 2019). Early diagnosis of leprosy is important to prevent the nerve damage in later stages, therefore, in 2016, the researchers identified it as the problem of the identification of lesions of leprosy as an imaging concern and deploys state-of-the-art architecture from the CNN project to address it by using DermnetNz datasets and achieved 91.6% accuracy of recognizing lesions (Baweja and Parhar 2016). Similarly, in 2018, scientists analyzed the epidemiology of leprosy by using the Kohonen Self-Organizing Maps algorithm to assess data from patients and their household contacts using Artificial Intelligence techniques. The findings examined illustrate a high number of late diagnoses and the values observed for the Anti PGL-1 in clusters suggesting a heavy leprosy bacillus burden and thus a high risk of contagion (da Silva et al. 2018). The Novartis Foundation and Microsoft have also collaborated to build an AI based digital tool enabling the early identification of leprosy (Novartis 2020). Irrespective of finding drug targets and diagnostics, AI is also used to find out SNPs and mutations to accurately define the types and lineages of the disease, as well as stability of the targeted proteins. A group of scientists recently used AI-based ML approaches to predict resistance in rpoB, inhA, katG, pncA, gyrA and gyrB genes for rifampicin, isoniazid, pyrazinamide, and fluoroquinolones (Jamal et al. 2020). In the construction of prediction models, they have used ML algorithms-naive bays, k nearest neighbor, support of the vector machine, and artificial neural network. The classification models had an overall precision of 85% for all genes tested and were evaluated for implementation using multiple unreported datasets (Jamal et al. 2020). These examples clearly illustrate that AI-based ML provides simple methods for complex research problems of prioritizing research compounds, which can also be used in diagnostics as well as to classify active molecules in accordance with medicinal chemistry insights.

9.4.2 AI and ML in Skin Diseases

Dermatology is the branch of medicine that treats the skin and its disorders. The causes of skin disorders include fungal, bacterial, allergic, and even insect bite disorders (Burns et al. 2008). They can also occur due to other diseases or because of the environment. Genetic factors also play a major role in the onset of a skin condition. Warts, Insect Bites, Psoriasis, Eczema, Meningitis, Measles, Ichthyosis, Acne, Scarlet Fever, and Stings are some examples of skin diseases (Hay et al. 2006). Erythematoscuamous class is one of the groups of skin diseases showing symptoms like the redness of the skin (erythema) is characterized by cell loss (squamous) (Azar et al. 2013). Psoriasis, seborrheic dermatitis, pityriasis rosea, chronic dermatitis, and lichen planus are some of the diseases that fall under the category of Erythematoscuamous class. It is very difficult to find out the specific illnesses that occur in a patient while diagnosing a skin disease, particularly of the groups of erythemato-scuamous diseases, the most common diseases in dermatology. Many researchers have tried to build automated systems that can predict this field. The artificial intelligence domain includes various algorithms, which are suited to developing diagnostic systems for skin diseases. Various examples are given hereby. In 2017, Esteva et al. published a seminal study in Nature that was noteworthy for being the first to compare the performance of a neural network with dermatologists in diagnosing skin cancer (Esteva et al. 2017). They used pre-trained GoogLeNet Inception v3 architecture and fine-tuned the network by using a dataset of 127,463 clinical and dermoscopic skin lesion images. Two hundred and sixty-five clinical images and 111 dermoscopic images of a ‘keratinocytic’ or ‘melanocytic’ type were provided to dermatologists and asked if they would: (1) prescribe biopsy or further care, or (2) reassure the patient. As a result, the average dermatologist was adequately recommending at a level below the CNN. Recently, the deep neural network algorithm was used by researchers for classifying dermoscopic images of four different skin diseases. The accuracy of Dataset A (1067 images) is 87.25 ± 2.24% and the accuracy of dataset B (528 similarly distributed) is 86.63% ± 5.78%. These four cutaneous diseases were Basal Cell Carcinoma (BCC), melanocytic nevus, sebourrheic keratosis (SK), and psoriasis (Zhang et al. 2018b). It is worth noting that the treatment of these diseases are completely different and, incorrect or delayed diagnosis may result in inappropriate care, delayed treatment, and even leads to death. It is also important for doctors to diagnose correctly in due course. After these four diseases automatically can be identified using the Artificial Intelligence System, clinicians can surely support patients by better and accurate diagnosis. In another report, 16,114 de-identified cases (photographs and clinical data) were used as differential diagnosis of skin conditions using a Deep learning for teledermatology practices. The DL differentiates between 26 common conditions of the skin, representing 80% of primary health cases and also claasifies 419 conditions of the skin. F 963 cases tested, the DL algorithm was not inferior to six other dermatologists and was higher than the six primary care physicians (PCPs) and six nurses (NPs) in the rotary panel of three board certified dermatologists (Liu et al. 2020b). In another landmark research, images from various websites related to different skin diseases have been collected. The database formed contains 80 photos of three diseases (20 Regular photographs, 20 photos of Melanoma, 20 images of eczema, and 20 images of psoriasis), and the method of detection was established with a pretrained, convolutionary neural system (AlexNet) and SVM that with 100% accuracy, the device successfully detects three different forms of skin disease. This approach takes a digital picture of the skin region of the disease effect, then uses image analyses to classify the disease type. It is easy and needs no costly equipment but a camera and a computer (ALEnezi and Method 2019). Skin disease identification constitutes a key step in reducing death rates, disease transmission, and skin disease growth. At present, treatment of these diseases are very costly and processed through a time-consuming clinical procedures. Artificial intelligence enables the development of automated dermatological screening techniques at an initial level, by focusing on image extraction, which is an important factor in the classification of skin diseases.

9.5 Limitations of AI and ML

The development of AI algorithms, the emergence of big data systems and the specialization of architectural hardware have contributed to the rapid growth of AI technology, especially in terms of the ML and DL approaches, alongside the development of the architectural hardware specialization, such as CPU, GPU, TPU, as well as large scale parallel computing (Yang et al. 2019). In several ways, AI has outpaced performance-related human experts. Therefore it is not shocking but exciting to use AI in drug research in a market along with a conservative approach (Miller and Brown 2018). AI is now coupled into the majority of pharmaceutical drug discovery phase, including problem recognition, hit/lead analysis, lead optimization, pharmacokinetic properties, toxicology, and clinical trial protocols (Fleming 2018). In spite of high boom during its inception, many obstacles are maintaining a calm head for AI applications in drug development. The collection of appropriate, high-quality, problem-specific data in particular remains a major challenge for the development of AI-assisted medicines (Yang et al. 2019; Fleming 2018). This is, sadly, simply not the case in the field of drug research, and there are many explanations why the standard or the quantity of data is not great. For one, confidence in the etiquette of data points depends highly on experimental circumstances, because of the extremely complicated biological structures under which medicines work (Yang et al. 2019). Various experimental conditions typically yield different or even contradictory effects. In contrast to the large amount of knowledge available to us, the amount of data available to us in the field of drug development is very limited (Jackson et al. 2018). Thus, the world needs not only the revolution in the process but also a revolution in the AI-assisted field of drug discovery (Fleming 2018; Sellwood et al. 2018; Zhong et al. 2018; Zhu 2020). A computer screening that is powered by machine learning is the next important constraint. Due to the difference of positive and negative results, current high-performance statistical approaches have the same issue as their theoretical equivalent (Gimeno et al. 2019). Moreover, in addition to the acceptance into clinical practice, interpretative performance is critical for revealing the information discovered by AI systems and for the identification of biases which may result in inappropriate behavior. In order to distinguish between bias, AI systems must be carefully implemented (Oliveira 2019; Fleming 2018; Dias and Torkamani 2019). When medical AI systems are not checked for distortion, they can function as disparity propagators. For example, DeepGestalt, an AI program for the study of facial dysmorphology, showed low precision in individuals of African versus European ancestry in defining the Down syndrome (36.8% vs. 80%, respectively) (Lumaka et al. 2017). The retraining of the Down syndrome model for African origin individuals has raised the Down syndrome diagnosis to 94.7%. Risk estimation in different population groups is also vulnerable to unequal output as training data under-representation (Martin et al. 2019). Nonetheless, tools are being developed which contribute to resolving the machine bias, which could not only help to overcome machine bias problems but also lead to diagnostic systems free of human bias (Chen et al. 2019). Profound learning can make maximum use of receptor, ligand information and their known interactions to help share knowledge from several studies and multiple targets to enhance our target performance. Researchers are expecting the huge boom in advancement of virtual screening technologies in the coming years to substitute or enhance the conventional high-performance screening process to increase the screening speed and success rate as the FDA has licensed growing numbers of AI algorithms (Topol 2019). However, these algorithms present a range of legal and ethical issues relating to data collection and privacy in the design and generalization of algorithms; for example, the legal procedure for updating this algorithms with new data and the responsibility of the prediction mistakes have not been touched yet (Topol 2019; Vayena et al. 2018). Providing an open source of AI models including the source codes, metagraphs, etc. to improve transparency could benefit the scientific and medical community (Dias and Torkamani 2019).

9.6 Can Machines Become a Total Replacement for Human Intelligence?

The concept of machines that overcome people can be connected inherently to conscious machines. Overcoming humans means replicating, meeting, and exceeding the main characteristics of human beings, such as high levels of consciousness (Signorelli 2018). Can computers be linked to humans, however? Could computers be aware? Could computers surpass the capability of humans? Those are paradoxical and contentious topics, in particular, because the knowledge of the brain is still secret and misunderstood. "Computing Machinery and Intelligence" is a landmark paper written by Alan Turing on the subject of artificial intelligence. The paper, published in Mind, in 1950, was the first to present to the general public his definition of what is now known as the Turing test (Turing 2009). Turing's paper answers the question "Can computers think?" Turing devised a test to address the question, in which computers held conversations with human judges. If the written answers of the computer foiled the judges into believing that he was a human, it could be assumed that it was a thinking machine. Though, human intelligence is quite unbelievable for all its faults. Without a doubt, scientists and businessmen enthusiasts did everything they could to replicate this in the form of artificial intelligence for over 60 years. While many reject such technology as the prelate of the future, it has enabled and even obsoleted countless activities. Many of the world's best minds work to develop artificial intelligence. The simplest example of this is playing chess on the computer. Computers are excellent in figuring out the next move in a game like chess, as the rules and patterns of the game have been well established but they need to communicate with the outside world, such as face recognition or understand spoken language that allows computers to manage variables that are constantly evolving and difficult to predict (Frankish and Ramsey 2014). The challenge with AI is that, however, many agree that it is a long way, if not impossible, to develop a program that can pass as human, not to say a rival of our mind. This has come a long way for artificial intelligence. Their ability to learn vast quantities of data, identify trends, and distribute results has improved numerous industries. Nevertheless, its greatest strength lies in the question of achieving true artificial intelligence: that it can't learn like a human. Human intelligence functions naturally and by incorporating various cognitive mechanisms to make up a certain view. Artificial intelligence, on the other hand, creates a model that can comfort like people, which seems unlikely, because nothing can replace a person with an artificial object. Biologically, for various reasons, the brain easily maintains the current intelligence lead on machines (Strukov et al. 2019). First of all, the information can be stored and processed within the same units, neurons and synapses. Secondly, in addition to superior architectural design, if neurons are taken for the comparative function, the brain has the advantage in cores number. Up to ten million cores are provided in advanced supercomputers, while the brain has almost 100 billion neurons (Oprea 2020). Nonetheless, the AI technique that currently drives virtually every area is linked to people's lives. In certain fields of study and education, AI is unavoidable. The rate of that is just picking up. This transition needs to be adapted and embraced by the human population.

9.7 Concluding Remarks

Many developments in the fields of physics, computer science, materials science, biology, genomics, and proteomics have been identified over the last decade. Such subtle yet disruptive innovations have unprecedentedly revolutionized medical practice as well as research outputs. Artificial intelligence equipped with ML and DL algorithms, biotechnological advances, such as precision genome editing, genomics, metabolomics and proteomics, and 'big data' would transform the understanding of the disease, its interpretations and patient supervision, and clinical data management. Such "Big Data" would make biological data more "holistic" because the artificial intelligence will consider several variables ranging from the genomics, metagenomics in real-time, to pathway interactions without violating the bias. This is significant from both the viewpoint of personalized medicine and public health through extreme modeling and simulation due to its 'predictive and preventive' capabilities. Machines are becoming increasingly effective in identifying and analyzing/diagnosing the many subtle signs that our bodies are misbehaving and, more significantly, in systematically researching and diagnosing diseases—they are on the road to excelling human beings. Slowly, as the technology progresses, they can be put to more general use, leading to lower medical expenditure. The emerging technology would allow the machines to manage and compare large quantities of data from multiple sources. Previously, machines may be constrained by their inputs, but currently, they have started enabling themselves to acquire inputs from multi-level genomic data that will surpass chemical sensors, human senses, physical senses, social context data, and 'big data' from genomics, proteomics, metabolomics to generate the significant output. Machines can process these data more efficiently than humans, resulting in quicker decision-making, better diagnosis and personalized patient care. Learning algorithms are rapidly improving the speed and efficiency of biological research as well as innovating the aspects of machine learning, such as conceptualization, generation of hypotheses, and even creativity that will ultimately be superior to humans. The existing artificial intelligence systems are, however, little more than a tool for helping the clinician develop the diagnosis and prediction. Today, the expertise of clinicians and scientists cannot be replicated by any algorithm, and it will take several years to combine or substitute human abilities and experiences with software performance altogether. Nonetheless, the potential for health care transformation in low- and middle-income countries, plagued by these infections, lies with the image-based artificial intelligence used in diagnosing neglected tropical diseases. Although this topic remains in its early stages, the clinical and public health environments in the most underserved areas should provide reliable diagnostic instruments.

Acknowledgments The authors thank Dr. Aparup Das, Director, ICMR-National Institute of Research in Tribal Health, Jabalpur for the encouragement and kind support. The manuscript has been approved by the Publication Screening Committee of ICMR-NIRTH, Jabalpur and assigned with the number ICMR-NIRTH/PSC/51/2020.

References

Advani J et al (2019) Whole genome sequencing of Mycobacterium tuberculosis clinical isolates from India reveals genetic heterogeneity and region-specific variations that might affect drug susceptibility. Front Microbiol 10:309

Akama T et al (2009) Whole-genome tiling array analysis of Mycobacterium leprae RNA reveals high expression of pseudogenes and noncoding regions. J Bacteriol 191(10):3321–3327

ALEnezi NSA, Method A (2019) Of skin disease detection using image processing and machine learning. Procedia Comput Sci 163:85–92

Alimadadi A et al (2020) Artificial intelligence and machine learning to fight COVID-19. American Physiological Society, Bethesda, MD

Allam Z (2020) The triple B: big data, biotechnology, and biomimicry. In: Biotechnology and future cities. Springer, Cham, pp 17–33

Arenas NE et al (2011) Molecular modeling and in silico characterization of Mycobacterium tuberculosis TlyA: possible misannotation of this tubercle bacilli-hemolysin. BMC Struct Biol 11(1):16

Azar AT et al (2013) Linguistic hedges fuzzy feature selection for differential diagnosis of Erythemato-squamous diseases. In: Soft computing applications. Springer, Berlin, pp 487–500

Bacher U et al (2018) Challenges in the introduction of next-generation sequencing (NGS) for diagnostics of myeloid malignancies into clinical routine use. Blood Cancer J 8(11):1–10

Bapat PM, Wangikar PP (2004) Optimization of rifamycin B fermentation in shake flasks via a machine-learning-based approach. Biotechnol Bioeng 86(2):201–208

Baweja HS, Parhar T (2016) Leprosy lesion recognition using convolutional neural networks. In: 2016 international conference on machine learning and cybernetics (ICMLC). IEEE

Belone AdFF et al (2015) Genome-wide screening of mRNA expression in leprosy patients. Front Genet 6:334

Benjak A et al (2018) Phylogenomics and antimicrobial resistance of the leprosy bacillus Mycobacterium leprae. Nat Commun 9(1):352

Bhandari J, Awais M, Gupta V (2020) Leprosy (Hansen Disease). In: StatPearls [internet]. StatPearls, Treasure Island, FL

Bleharski JR et al (2003) Use of genetic profiling in leprosy to discriminate clinical forms of the disease. Science 301(5639):1527–1530

Burns T et al (2008) Rook's textbook of dermatology. Wiley, Hoboken, NJ
Buyuksimsek M et al (2019) Results of liquid biopsy studies by next generation sequencing in patients with advanced stage non-small cell lung cancer: single center experience from Turkey. Balkan J Med Genet 22(2):17–24
Camacho DM et al (2018) Next-generation machine learning for biological networks. Cell 173 (7):1581–1592
Chakrabarty S et al (2019) Host and MTB genome encoded miRNA markers for diagnosis of tuberculosis. Tuberculosis 116:37–43
Chance MR et al (2004) High-throughput computational and experimental techniques in structural genomics. Genome Res 14(10b):2145–2154
Chatterjee A et al (2017) Whole genome sequencing of clinical strains of Mycobacterium tuberculosis from Mumbai, India: a potential tool for determining drug-resistance and strain lineage. Tuberculosis 107:63–72
Chatterjee S, Poonawala H, Jain Y (2018) Drug-resistant tuberculosis: is India ready for the challenge? BMJ Glob Health 3(4):e000971
Chen IY, Szolovits P, Ghassemi M (2019) Can AI help reduce disparities in general medical and mental health care? AMA J Ethics 21(2):167–179
Chetty S et al (2017) Recent advancements in the development of anti-tuberculosis drugs. Bioorg Med Chem Lett 27(3):370–386
Cole ST, Supply P, Honore N (2001) Repetitive sequences in Mycobacterium leprae and their impact on genome plasticity. Lepr Rev 72(4):449–461
da Silva YED et al (2018) Application of clustering technique with Kohonen self-organizing maps for the epidemiological analysis of leprosy. In: Proceedings of SAI intelligent systems conference. Springer, Berlin
Dagiasis AP et al (2014) A high performance biomarker detection Method for exhaled breath mass spectrometry data. In: Topics in nonparametric statistics. Springer, Cham, pp 207–216
de Souza GA et al (2009) Validating divergent ORF annotation of the Mycobacterium leprae genome through a full translation data set and peptide identification by tandem mass spectrometry. Proteomics 9(12):3233–3243
Deepika K, Seema S (2016) Predictive analytics to prevent and control chronic diseases. In: 2016 2nd international conference on applied and theoretical computing and communication technology (iCATccT). IEEE
Denton JF et al (2014) Extensive error in the number of genes inferred from draft genome assemblies. PLoS Comput Biol 10(12):e1003998
Di Resta C et al (2018) Next-generation sequencing approach for the diagnosis of human diseases: open challenges and new opportunities. Ejifcc 29(1):4
Dias R, Torkamani A (2019) Artificial intelligence in clinical and genomic diagnostics. Genome Med 11(1):1–12
Djaout K et al (2016) Predictive modeling targets thymidylate synthase ThyX in Mycobacterium tuberculosis. Sci Rep 6(1):1–11
Dorhoi A et al (2013) MicroRNA-223 controls susceptibility to tuberculosis by regulating lung neutrophil recruitment. J Clin Invest 123(11):4836–4848
Ekins S et al (2014) Bayesian models for screening and TB Mobile for target inference with Mycobacterium tuberculosis. Tuberculosis 94(2):162–169
Eloit M (2014) The diagnosis of infectious diseases by whole genome next generation sequencing: a new era is opening. Front Cell Infect Microbiol 4:25
Esfandyarpour R et al (2013) Simulation and fabrication of a new novel 3D injectable biosensor for high throughput genomics and proteomics in a lab-on-a-chip device. Nanotechnology 24 (46):465301
Esteva A et al (2017) Dermatologist-level classification of skin cancer with deep neural networks. Nature 542(7639):115–118
Fauci AS, Lane HC, Redfield RR (2020) Covid-19—navigating the uncharted. N Engl J Med 382 (13):1268–1269

Fleming N (2018) How artificial intelligence is changing drug discovery. Nature 557(7706):S55–S55

Frankish K, Ramsey WM (2014) The Cambridge handbook of artificial intelligence. Cambridge University Press, Cambridge

Gama RS et al (2019) A novel integrated molecular and serological analysis method to predict new cases of leprosy amongst household contacts. PLoS Negl Trop Dis 13(6):e0007400

Gimeno A et al (2019) The light and dark sides of virtual screening: what is there to know? Int J Mol Sci 20(6):1375

Grossman SR et al (2013) Identifying recent adaptations in large-scale genomic data. Cell 152 (4):703–713

Guigó R et al (2006) EGASP: the human ENCODE genome annotation assessment project. Genome Biol 7(S1):S2

Gupta AK, Gupta U (2014) Next generation sequencing and its applications. In: Animal biotechnology. Elsevier, Amsterdam, pp 345–367

Harrow J et al (2009) Identifying protein-coding genes in genomic sequences. Genome Biol 10 (1):201

Hay R et al (2006) Skin diseases. In: Disease control priorities in developing countries, 2nd edn. The International Bank for Reconstruction and Development/The World Bank, Washington, DC

Hodgson DR, Wellings R, Harbron C (2012) Practical perspectives of personalized healthcare in oncology. New Biotechnol 29(6):656–664

Hodkinson BP, Grice EA (2015) Next-generation sequencing: a review of technologies and tools for wound microbiome research. Adv Wound Care 4(1):50–58

Hu X et al (2020) LncRNA and predictive model to improve the diagnosis of clinically diagnosed pulmonary tuberculosis. J Clin Microbiol 58:e01973-19

Ioerger TR et al (2013) Identification of new drug targets and resistance mechanisms in Mycobacterium tuberculosis. PLoS One 8(9):e75245

Islam MM et al (2017) Drug resistance mechanisms and novel drug targets for tuberculosis therapy. J Genet Genomics 44(1):21–37

Israel21c (2020) Covid spit test. https://www.israel21c.org/israeli-1-second-covid-spit-test-shown-95-accurate-so-far/

Jackson N, Czaplewski L, Piddock LJ (2018) Discovery and development of new antibacterial drugs: learning from experience? J Antimicrob Chemother 73(6):1452–1459

Jamal S et al (2020) Artificial intelligence and machine learning based prediction of resistant and susceptible mutations in Mycobacterium tuberculosis. Sci Rep 10(1):1–16

Jiang H, He K (2020) Statistics in the Genomic Era. Multidisciplinary Digital Publishing Institute, Basel

Joshi RS et al (2013) Resistome analysis of Mycobacterium tuberculosis: identification of aminoglycoside 2'-Nacetyltransferase (AAC) as co-target for drug desigining. Bioinformation 9(4):174

Kalo D et al (2015) Pattern of drug resistance of Mycobacterium tuberculosis clinical isolates to first-line antituberculosis drugs in pulmonary cases. Lung India 32(4):339

Kim K et al (2020) Predicting clinical benefit of immunotherapy by antigenic or functional mutations affecting tumour immunogenicity. Nat Commun 11(1):1–11

Kinnings SL et al (2011) A machine learning-based method to improve docking scoring functions and its application to drug repurposing. J Chem Inf Model 51(2):408–419

Kouchaki S et al (2019) Application of machine learning techniques to tuberculosis drug resistance analysis. Bioinformatics 35(13):2276–2282

Koumakis L (2020) Deep learning models in genomics; are we there yet? Comput Struct Biotechnol J 18:1466–1473

Kumar K, Abubakar I (2015) Clinical implications of the global multidrug-resistant tuberculosis epidemic. Clin Med 15(Sup 6):s37–s42

Kwan PKW et al (2020) Gene expression responses to anti-tuberculous drugs in a whole blood model. BMC Microbiol 20:1–9
Lane T et al (2018) Comparing and validating machine learning models for mycobacterium tuberculosis drug discovery. Mol Pharm 15(10):4346–4360
Lavania M et al (2015) Genotyping of Mycobacterium leprae strains from a region of high endemic leprosy prevalence in India. Infect Genet Evol 36:256–261
Lavania M et al (2018) Molecular detection of multidrug-resistant Mycobacterium leprae from Indian leprosy patients. J Glob Antimicrob Resist 12:214–219
Lebrigand K et al (2020) High throughput error corrected Nanopore single cell transcriptome sequencing. Nat Commun 11(1):1–8
Li L et al (2020) Artificial intelligence distinguishes COVID-19 from community acquired pneumonia on chest CT. Radiology 296(2):200905
Libbrecht MW, Noble WS (2015) Machine learning applications in genetics and genomics. Nat Rev Genet 16(6):321–332
Liu C et al (2013) Applications of machine learning in genomics and systems biology. Comput Math Methods Med 2013:587492
Liu H et al (2020a) A panel of circRNAs in the serum serves as biomarkers for mycobacterium tuberculosis infection. Front Microbiol 11:1215
Liu Y et al (2020b) A deep learning system for differential diagnosis of skin diseases. Nat Med 26:900–908
Lohiya A et al (2020) Prevalence and patterns of drug resistant pulmonary tuberculosis in India—a systematic review and meta-analysis. J Glob Antimicrob Resist 22:308–316
Lumaka A et al (2017) Facial dysmorphism is influenced by ethnic background of the patient and of the evaluator. Clin Genet 92(2):166–171
Lv W et al (2020) Discovery and validation of biomarkers for Zhongning goji berries using liquid chromatography mass spectrometry. J Chromatogr B 1142:122037
Lyko K, Nitzschke M, Ngomo A-CN (2016) Big data acquisition. In: New horizons for a data-driven economy. Springer, Cham, pp 39–61
Mak K-K, Pichika MR (2019) Artificial intelligence in drug development: present status and future prospects. Drug Discov Today 24(3):773–780
Manson AL et al (2017) Mycobacterium tuberculosis whole genome sequences from southern India suggest novel resistance mechanisms and the need for region-specific diagnostics. Clin Infect Dis 64(11):1494–1501
Manzoni C et al (2018) Genome, transcriptome and proteome: the rise of omics data and their integration in biomedical sciences. Brief Bioinform 19(2):286–302
Marques MAM et al (2008) Deciphering the proteomic profile of Mycobacterium leprae cell envelope. Proteomics 8(12):2477–2491
Martin AR et al (2019) Clinical use of current polygenic risk scores may exacerbate health disparities. Nat Genet 51(4):584–591
Matsuoka M (2010) Drug resistance in leprosy. Jpn J Infect Dis 63(1):1–7
Matsuoka M et al (2007) The frequency of drug resistance mutations in Mycobacterium leprae isolates in untreated and relapsed leprosy patients from Myanmar, Indonesia and the Philippines. Lepr Rev 78(4):343–352
McNerney R, Zignol M, Clark TG (2018) Use of whole genome sequencing in surveillance of drug resistant tuberculosis. Expert Rev Anti-Infect Ther 16(5):433–442
Mehaffy C et al (2018) Biochemical characterization of isoniazid-resistant Mycobacterium tuberculosis: can the analysis of clonal strains reveal novel targetable pathways? Mol Cell Proteomics 17(9):1685–1701
Mehta MD, Liu PT (2014) microRNAs in mycobacterial disease: friend or foe? Front Genet 5:231
Mertz L (2020) AI-driven COVID-19 tools to interpret, quantify lung images. IEEE Pulse 11 (4):2–7
Miller DD, Brown EW (2018) Artificial intelligence in medical practice: the question to the answer? Am J Med 131(2):129–133

Mishra R et al (2020) Potential role of adjuvant drugs on efficacy of first line oral antitubercular therapy: drug repurposing. Tuberculosis 120:101902

Mohanty S et al (2020) Application of artificial intelligence in COVID-19 drug repurposing. Diabetes Metab Syndr Clin Res Rev 14(5):1027–1031

Mokrousov I et al (2016) Next-generation sequencing of Mycobacterium tuberculosis. Emerg Infect Dis 22(6):1127

Monot M et al (2009) Comparative genomic and phylogeographic analysis of Mycobacterium leprae. Nat Genet 41(12):1282–1289

Munir K et al (2019) Cancer diagnosis using deep learning: a bibliographic review. Cancers 11 (9):1235

Nagpal P et al (2020) Long-range replica exchange molecular dynamics guided drug repurposing against tyrosine kinase PtkA of Mycobacterium tuberculosis. Sci Rep 10(1):1–11

Nobre T et al (2016) Misannotation awareness: a tale of two gene-groups. Front Plant Sci 7:868

Novartis (2020) AI-powered diagnostic tool to aid in the early detection of leprosy. https://www.novartisfoundation.org/news/ai-powered-diagnostic-tool-aid-early-detection-leprosy

Oliveira AL (2019) Biotechnology, big data and artificial intelligence. Biotechnol J 14(8):1800613

Oprea R (2020) AI versus the human brain. Brain Minds. https://brandminds.live/

Oprea T, Mestres J (2012) Drug repurposing: far beyond new targets for old drugs. AAPS J 14 (4):759–763

Pan S-Y et al (2014) Historical perspective of traditional indigenous medical practices: the current renaissance and conservation of herbal resources. Evid Based Complement Alternat Med 2014:525340

Papadopoulou E et al (2019) Clinical feasibility of NGS liquid biopsy analysis in NSCLC patients. PLoS One 14(12):e0226853

Paranjpe MD, Taubes A, Sirota M (2019) Insights into computational drug repurposing for neurodegenerative disease. Trends Pharmacol Sci 40(8):565–576

Parkash O, Singh B (2012) Advances in proteomics of Mycobacterium leprae. Scand J Immunol 75 (4):369–378

Parvathaneni V et al (2019) Drug repurposing: a promising tool to accelerate the drug discovery process. Drug Discov Today 24(10):2076–2085

Passi A et al (2018) RepTB: a gene ontology based drug repurposing approach for tuberculosis. J Chem 10(1):24

Pauli I et al (2013) Discovery of new inhibitors of Mycobacterium tuberculosis InhA enzyme using virtual screening and a 3D-pharmacophore-based approach. J Chem Inf Model 53 (9):2390–2401

Pedlar CR, Newell J, Lewis NA (2019) Blood biomarker profiling and monitoring for high-performance physiology and nutrition: current perspectives, limitations and recommendations. Sports Med 49(2):185–198

Peng Z, Chen L, Zhang H (2020) Serum proteomic analysis of Mycobacterium tuberculosis antigens for discriminating active tuberculosis from latent infection. J Int Med Res 48 (3):0300060520910042

Pinto SM et al (2018) Integrated multi-omic analysis of Mycobacterium tuberculosis H37Ra redefines virulence attributes. Front Microbiol 9:1314

Prada CF, Boore JL (2019) Gene annotation errors are common in the mammalian mitochondrial genomes database. BMC Genomics 20(1):73

Prathipati P, Ma NL, Keller TH (2008) Global Bayesian models for the prioritization of antitubercular agents. J Chem Inf Model 48(12):2362–2370

Priya Doss CG et al (2014) Integrating in silico prediction methods, molecular docking, and molecular dynamics simulation to predict the impact of ALK missense mutations in structural perspective. Biomed Res Int 2014:895831

Pushkaran AC et al (2019) Combination of repurposed drug diosmin with amoxicillin-clavulanic acid causes synergistic inhibition of mycobacterial growth. Sci Rep 9(1):1–14

Pushpakom S et al (2019) Drug repurposing: progress, challenges and recommendations. Nat Rev Drug Discov 18(1):41–58
Qi Y et al (2012) Altered serum microRNAs as biomarkers for the early diagnosis of pulmonary tuberculosis infection. BMC Infect Dis 12(1):384
Qin D (2019) Next-generation sequencing and its clinical application. Cancer Biol Med 16(1):4
Rajkomar A, Dean J, Kohane I (2019) Machine learning in medicine. N Engl J Med 380 (14):1347–1358
Rani J et al (2020) Repurposing of FDA-approved drugs to target MurB and MurE enzymes in Mycobacterium tuberculosis. J Biomol Struct Dyn 38(9):2521–2532
Rao PN, Suneetha S (2018) Current situation of leprosy in India and its future implications. Indian Dermatol Online J 9(2):83
Rao ASS, Vazquez JA (2020) Identification of COVID-19 can be quicker through artificial intelligence framework using a mobile phone–based survey when cities and towns are under quarantine. Infect Control Hosp Epidemiol 41(7):826–830
Romanowski K et al (2020) Using whole genome sequencing to determine the timing of secondary tuberculosis in British Columbia, Canada. Clin Infect Dis. https://doi.org/10.1093/cid/ciaa1224
Rufai SB, Singh S (2019) Whole-genome sequencing of two extensively drug-resistant Mycobacterium tuberculosis isolates from India. Microbiol Resour Announc 8(7):e00007-19
Sarnaik A et al (2020) High-throughput screening for efficient microbial biotechnology. Curr Opin Biotechnol 64:141–150
Schneider G (2018) Automating drug discovery. Nat Rev Drug Discov 17(2):97
Schuenemann VJ et al (2013) Genome-wide comparison of medieval and modern Mycobacterium leprae. Science 341(6142):179–183
Scollard DM et al (2006) The continuing challenges of leprosy. Clin Microbiol Rev 19(2):338–381
Sellwood MA et al (2018) Artificial intelligence in drug discovery. Future Med Chem 10 (17):2025–2028. https://doi.org/10.4155/fmc-2018-0212
Shi F et al (2020) Review of artificial intelligence techniques in imaging data acquisition, segmentation and diagnosis for covid-19. IEEE Rev Biomed Eng 14:4–15
Signorelli CM (2018) Can computers overcome humans? Consciousness interaction and its implications. In: 2018 IEEE 17th international conference on cognitive informatics & cognitive computing (ICCI* CC). IEEE
Silva DR et al (2018) New and repurposed drugs to treat multidrug-and extensively drug-resistant tuberculosis. J Bras Pneumol 44(2):153–160
Singh P, Cole ST (2011) Mycobacterium leprae: genes, pseudogenes and genetic diversity. Future Microbiol 6(1):57–71
Singh P et al (2015) Insight into the evolution and origin of leprosy bacilli from the genome sequence of Mycobacterium lepromatosis. Proc Natl Acad Sci 112(14):4459–4464
Singh A, Somvanshi P, Grover A (2019) Drug repurposing against arabinosyl transferase (EmbC) of Mycobacterium tuberculosis: essential dynamics and free energy minima based binding mechanics analysis. Gene 693:114–126
Singh R et al (2020) Recent updates on drug resistance in Mycobacterium tuberculosis. J Appl Microbiol 128(6):1547–1567
Spinelli SV et al (2013) Altered microRNA expression levels in mononuclear cells of patients with pulmonary and pleural tuberculosis and their relation with components of the immune response. Mol Immunol 53(3):265–269
Steinegger M, Salzberg SL (2020) Terminating contamination: large-scale search identifies more than 2,000,000 contaminated entries in GenBank. Genome Biol 21(1):1–12
Strukov D et al (2019) Building brain-inspired computing. Nat Commun 10(1):4838
Štular T et al (2016) Discovery of mycobacterium tuberculosis InhA inhibitors by binding sites comparison and ligands prediction. J Med Chem 59(24):11069–11078
Sun W et al (2016) Rapid antimicrobial susceptibility test for identification of new therapeutics and drug combinations against multidrug-resistant bacteria. Emerg Microb Infect 5(1):1–11

Swayamsiddha S, Mohanty C (2020) Application of cognitive internet of medical things for COVID-19 pandemic. Diabetes Metab Syndr Clin Res Rev 14(5):911–915

Tagliani E et al (2021) Use of a whole genome sequencing-based approach for Mycobacterium tuberculosis surveillance in Europe in 2017–2019: an ECDC pilot study. Eur Respir J 57:2002272

Thakur V, Varshney R (2010) Challenges and strategies for next generation sequencing (NGS) data analysis. J Comput Sci Syst Biol 3:40–42

Ting DSW et al (2020) Digital technology and COVID-19. Nat Med 26(4):459–461

Tio-Coma M et al (2020) Detection of new Mycobacterium leprae subtype in Bangladesh by genomic characterization to explore transmission patterns. medRxiv

Topol EJ (2019) High-performance medicine: the convergence of human and artificial intelligence. Nat Med 25(1):44–56

Truman RW et al (2011) Probable zoonotic leprosy in the southern United States. N Engl J Med 364 (17):1626–1633

Turing AM (2009) Computing machinery and intelligence. In: Parsing the turing test. Springer, Dordrecht, pp 23–65

Uddin R et al (2016) Computational identification of potential drug targets against Mycobacterium leprae. Med Chem Res 25(3):473–481

Vaishya R et al (2020) Artificial intelligence (AI) applications for COVID-19 pandemic. Diabetes Metab Syndr Clin Res Rev 14(4):337–339

Vayena E, Blasimme A, Cohen IG (2018) Machine learning in medicine: addressing ethical challenges. PLoS Med 15(11):e1002689

Wadapurkar RM, Vyas R (2018) Computational analysis of next generation sequencing data and its applications in clinical oncology. Inform Med Unlocked 11:75–82

Wakeling MN et al (2019) Misannotation of multiple-nucleotide variants risks misdiagnosis. Wellcome Open Res 4:145

Waman VP et al (2019) Mycobacterial genomics and structural bioinformatics: opportunities and challenges in drug discovery. Emerg Microb Infect 8(1):109–118

Wan L et al (2020) Genomic analysis identifies mutations concerning drug-resistance and Beijing genotype in multidrug-resistant Mycobacterium tuberculosis isolated from China. Front Microbiol 11:1444

Wanichthanarak K, Fahrmann JF, Grapov D (2015) Genomic, proteomic, and metabolomic data integration strategies. Biomark Insights 10(Suppl 4):1–6

Williams DL et al (2009) Implications of high level pseudogene transcription in Mycobacterium leprae. BMC Genomics 10:397

Wilsey C et al (2013) A large scale virtual screen of DprE1. Comput Biol Chem 47:121–125

World Health Organization (2018) Global health TB report. WHO, Geneva

Wu J et al (2012) Analysis of microRNA expression profiling identifies miR-155 and miR-155* as potential diagnostic markers for active tuberculosis: a preliminary study. Hum Immunol 73 (1):31–37

Xia J, Benner MJ, Hancock RE (2014) NetworkAnalyst-integrative approaches for protein–protein interaction network analysis and visual exploration. Nucleic Acids Res 42(W1):W167–W174

Xiong Y et al (2018) Automatic detection of mycobacterium tuberculosis using artificial intelligence. J Thorac Dis 10(3):1936

Yang X et al (2019) Concepts of artificial intelligence for computer-assisted drug discovery. Chem Rev 119(18):10520–10594

Yi Z et al (2012) Altered microRNA signatures in sputum of patients with active pulmonary tuberculosis. PLoS One 7(8):e43184

Yohe S, Thyagarajan B (2017) Review of clinical next-generation sequencing. Arch Pathol Lab Med 141(11):1544–1557

Zak DE et al (2016) A blood RNA signature for tuberculosis disease risk: a prospective cohort study. Lancet 387(10035):2312–2322

Zhang W, Cheng B, Bingying X (2017) Application of next-generation sequencing technology in forensic science. Chin J Forensic Med 32(1):40–43
Zhang G et al (2018a) Virtual screening of small molecular inhibitors against DprE1. Molecules 23 (3):524
Zhang X et al (2018b) Towards improving diagnosis of skin diseases by combining deep neural network and human knowledge. BMC Med Inform Decis Mak 18(2):59
Zhang H et al (2019) NCNet: deep learning network models for predicting function of non-coding DNA. Front Genet 10:432
Zhong F et al (2018) Artificial intelligence in drug design. Sci China Life Sci 61(10):1191–1204
Zhu H (2020) Big data and artificial intelligence modeling for drug discovery. Annu Rev Pharmacol Toxicol 60:573–589
Zou J et al (2019) A primer on deep learning in genomics. Nat Genet 51(1):12–18
Zuniga ES, Early J, Parish T (2015) The future for early-stage tuberculosis drug discovery. Future Microbiol 10(2):217–229

Bias in Medical Big Data and Machine Learning Algorithms

10

Abstract

Data intensive technologies using medical big data, analysed by machine learning algorithms, play a key role in revolutionising healthcare. However, results from several findings show that these algorithms have potential to gain negative impact on healthcare system as compared to the existing primitive healthcare systems which involve physicians. Current algorithms are accused of these deficiencies resulting from biased training data bearing numerous missing values, errors, and biased inputs. This is due to under- or over-representation of certain groups of data, trivial data curation methods, etc. In this chapter, we describe Perceptive Bias, Processing Bias, and the ways to compute bias for Medical Big Data analysis.

Keywords

Artificial intelligence · Machine learning · Medical big data · Big data analytics · Algorithms · Bias · mHealth

10.1 Introduction

With evolution in humankind, creativity of our brain leads to invention of machines. Unlike the human brain, machines do not have ability to interpret the data and make decisions. It was not until mid-twentieth century, when Turing made the first Artificially Intelligent (AI) machine which had the ability to think. After the discovery of first Neural Network by Pitts and McCulloch in 1943, there was a revolution with a question: can a machine think? With recent advancements in technology, machines can now focus on vision, hearing, natural languages processing, image processing and pattern recognition, cognitive computing, knowledge representation, and many more. These findings helped Machine Learning (ML) acquire the ability to

A. Saxena, S. Chandra, *Artificial Intelligence and Machine Learning in Healthcare*, https://doi.org/10.1007/978-981-16-0811-7_10

generate a huge quantum of data through sensors, just like humans, and process it using computational intelligence (Skilling and Gull 1985). This huge quantity of data can be termed as Big data.

Big data can be defined as datasets which are so diverse and complex in scale that it cannot be managed and analysed by existing data base management systems and thus requires new architectural framework, algorithms for its management (Lee and Yoon 2017). Although Big data is characterised by its V's, i.e. Volume, Velocity, and Variety, which in itself represents its gigantic size and the tremendous values and knowledge hidden in it which could significantly benefit the Big data shareholders (Arora 2018). Smartphone's. Big data lately came into prominence because of data intensive technologies, as we are residing in the world which utilises enough amounts of data.

Big Data is basically categorised into three major types that is structured, semi-structured, and unstructured data. Structured-data concerns all data which is stored in the database in tabular form. Structured data represent only 5–10% of all informatics data. For example, relational data. Semi-Structured data is information that does not inhabit in a relational database but that does have some organisational properties that make it easier to analyse. For example, CSV sttructured and XML, JSON documents are semi structured documents, NoSQL databases, considered as Semi-Structured and Unstructured data, represent around 80% of data. Unstructured data is everywhere. In fact, most individuals and organisations achieve their lives around free data. For example, video-graphic documents, word-processing documents, photographic documents, presentations, webpages, and many other kinds of business documents, audio files, Electronic-mails, Word files, PDF's, Text's, Media Logs,... (Cirillo and Valencia 2019).

Big Data infrastructure is a framework which covers important components, including Hadoop (hadoop.apache.org), NoSQL databases, massively parallel processing (MPP), and others, that are used for storing, processing, and analysing Big Data. Big Data analytics covers collection, manipulation, and analyses of massive, diverse datasets that contain a variety of data types, including genomic data and EHRs to reveal hidden patterns, cryptic correlations, and other intuitions on a Big Data infrastructure (He et al. 2017).

In this chapter, we discuss about the sources of medical big data, machine learning, and artificial intelligence algorithms used to analyse the medical big data and the potential reasons of bias in the data which raise a question about use of machines without human intervention in healthcare. The most common reasons for bias during data curation could be corruption of data, redundant or missing records, missing values, etc., which, cumulatively, increases over the process of structuring, processing, and analysing which could result in false predictions.

10.2 Medical Big Data (MBD)

There are various sources of medical Big data not limiting to medical health records, electronic healthcare records, clinical registries, diagnostic reports, biometrics, patient reported data (mHealth), data over internet, diagnostic and medical imaging, genetic/molecular bio-markers, data from coherent studies, data from clinical trials, routine check-ups, and smart phone generated data in real time (He et al. 2017; Saxena and Saxena 2020; Savage 2012).

Integration of this medical big data from various sources cause complements the dimension of the data, which amplifies itself to multiple folds, thus becomes complex and incorporates redundancy, incompleteness, incongruence resulting in bias with cumulative increase over the successive levels. Medical big data (MBD) varies from Big Data from other disciples thus its generally hard to analyse and extract knowledge for most investigators, making practice of open datascience or medical Big Data Analytics (BDA) less popular due to ethical concerns, risk of misuse of data by third parties and unavailability of open source reliable data in public domain (Jensen 2018).

MBD is relatively new. Thus it is usually curated and collected using pre-defined protocol in fixed forms, thus they are relatively more structured than big data from other disciplines. This is mainly due to the well-structured data extraction process that simplifies the raw data (He et al. 2017; Denny et al. 2018). Curation of MBD is expensive due to involvement of skilled man-power, expensive instrumentation (diagnostic and imaging platforms, sensors, etc.), and especially due to involvement of human population as subjects (e.g., Clinical trials). Thus availability of MBD is relatively limited and is usually collected in non-reproducible situation, affected by various sources of uncertainty at each level (due to human involvement), such as missing data, measurement errors, technical collapses, etc. (Ntoutsi et al. 2020).

Potential applications of MBD can be found in personalised medicine, clinical decision support system, diagnostic and treatment decision to support patient's behaviour using mobile device, population health analysis, fraud detection and prevention, etc. (Denny et al. 2018; Ntoutsi et al. 2020) Based on these applications, Data analytics for MBD could be used in various healthcare sectors to improve quality of healthcare, including predictive modelling (for the optimum use of resources and accessing risks), management of population, surveillance of medical device safety and drugs, monitoring heterogeneity in treatment and disease, clinical decision support and personalised medicine, performance measurement, thus improving quality of care, monitoring public health and research applications (Rumsfeld et al. 2016).

10.3 Analysis of Medical Big Data

Data science algorithms enable machines to perform tasks skilfully, using artificial intelligence. They require data to learn, thus they require datasets to train themselves before predictive models can be obtained (Skilling and Gull 1985). There are several

ML algorithms used to analyse and predict MBD, such as Decision Tree (DT), Naïve Bayes (NB) classifiers, k-nearest neighbours (k-NN), Support Vector Machine (SVM), Artificial Neural Network (ANN), Deep Learning (DL), etc.

In Decision Tree (Ramírez et al. 2019), a simple algorithm creates mutually exclusive classes by answering questions in a predefined order. Naïve Bayes (NB) classifiers, output probabilistic dependencies among variables. In k-nearest neighbours, a feature classified according to its closest neighbour in the dataset, are used for classification and regression. Support Vector Machine uses a trained model which will classify new data into categories. It can find complex patterns by choosing kernels which perform transformation of data and choose support vectors. Artificial Neural Network is used to approximate functions. They have several layers of neuron resembling human. Each "neuron" has a weight that determines its importance. Each layer receives data from the previous layer, calculates a score, and passes the output to the next layer. It is considered supervised machine learning. Deep Learning uses a variant of ANNs, where multiple layers of neurons are used. It can perform both supervised or unsupervised learning (Tang et al. 2019; Bibault et al. 2016).

10.4 Bias

Bias is not a new problem, rather "Bias is as old as human civilization" and "it is human nature for members of the dominant majority to be oblivious to the experiences of other groups" (Jensen 2018; Saxena et al. 2021). Artificial Intelligence (supervised or unsupervised learning) algorithms are significantly employed in public and private domains to make decisions which are beyond the capabilities of human, which have long term impact on mankind and society. However, these algorithms may cumulatively amplify the pre-existing bias in MBD which, consciously or unconsciously, incurred during data curation and analysis thus evolving new criteria and classification with tremendous potential for new bias. This had led to increasing concern among data scientists and curators to reconsider the artificially intelligent system and its associated algorithms towards new approaches which efficiently solves the purpose with sensitivity addressing the fairness of the decision thus reducing the chances of bias.

For the ease of categorisation in this chapter, Bias is divided into three different classes (Fig. 10.1).

- *Perceptive Bias*
  These are the approaches which understand the origin or creation of bias in the society followed by its entry into the social and technical systems and ultimately manifestation or fairness of data used by AI algorithms, which can be defined formally and modelled to give a knowledgeable outcome.
- *Processing Bias*
  As the name suggests, this approach deals with the bias which pioneers during different stages of decision-making by AI algorithms primarily focusing on input

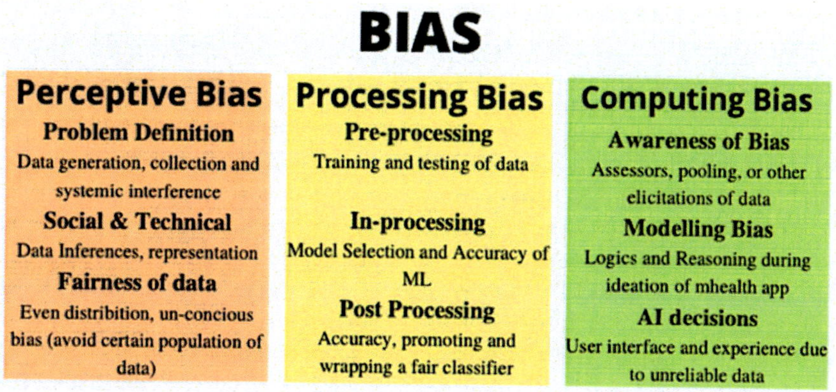

Fig. 10.1 Overview of Bias

of data by the user, training or learning of AI algorithms and model output during pre-processing, processing, and post processing, respectively.

- *Computing Bias*
  This includes approaches which account for pro-activity of bias throughout the process via retro-activity or bias aware data collection.

However, as the AI algorithms and AI technology crawls deep into the society, it is important for data scientists and algorithm creators to be aware of conscious, subconscious or unconscious discrimination or bias due to any past incident in life to ensure the responsible usage of technology, keeping in mind that "a technological approach on its own is not a panacea for all sort of bias and AI problems" (Ntoutsi et al. 2020).

10.4.1 Perceptive Bias

Bias is a primitive notion for Machine Learning and AI algorithms, which was trivially referred to assumptions or educated guesses made by specific model or curator themselves (Mitchell 1997). Stab et al. in their survey studied about "inclination or prejudice of a decision made by an AI system which is for or against one person or group, especially in a way considered to be unfair" (Ntoutsi et al. 2020). This survey supports our assumption about how bias enters the data analytics system and how it is incorporated as the part of data which serves as input data to AI algorithms. Further, we discuss various aspects of perceptive bias and their definition with example of mhealth- or smartphone-generated medical data.

10.4.1.1 Problem Definition

One of the major challenges with MBD curation is collection and its storage. There are no standard protocols set for the data curation yet, which could address the problems of missing values. Also data curation involves involvement of human manpower, which is a potential source of bias due to its perception and understanding. For instance, with advancements in technology and reducing cost of sensors and chip, smartphones or mhealth devices (smart gears and sensory wearables) are becoming more and more popular among masses contributing to larger proportions of MBD (Saxena and Saxena 2020). However, there are various parameters which are directly monitored and recorded by sensors, such as oxygen saturation, pulse rate, etc., whereas some parameters need interference and human validation, such as calories intake, sleep wake cycle, etc. This human intervention could be a potential source of bias due to lack of understanding between the user and AI interface, wrong inputs recorded (human manipulation to satisfy user's need), sensory failure, etc.

10.4.1.2 Social and Technical Aspects

As mentioned in the previous section, data analytics by AI depends directly upon the data collected from humans manually (trivial health records or via software created by humans (mHealth data). Thus the innate bias which exists in humans is acquired by the data analytical systems. And further, the bias in the data is amplified due to complex sociotechnical systems, resulting in inequalities and discrimination. This directly depends on the representation of data and how it has been inferred during the analysis process. Sometimes the algorithms may amplify or introduce bias to favour some component or aspects of human behaviour, thus shaping social institutions. However this is currently not clear and requires more scientific interventions.

Social bias can be introduced in data through sensitive features in the form of data values. Interdependence between the data in the dataset or simple co-relations between neutral features could potentially lead to bias. Representation of different strata of data in a dataset is another aspect to minimise technical bias. Machine learning (ML) algorithms and other statistical inferences require training models (training datasets) of data on which they are trained and applied. This generally leads to under- or over-representation of certain strata of data, especially for medical big data, as they are not curated primarily for these algorithms. Another parameter that needs to be taken into account is the structure of data. Generally ML applications work on structured data, whereas MBD is significantly unstructured and thus introduces bias in some strata or the other.

10.4.1.3 Fairness of Data

Data fairness could be defined as the fair representation of different groups of data at each stratum in the dataset considering predicted and actual outcomes, which certainly rely on demographic parity, equalised odds, and correct calibration.

For the large data sets representing certain strata of the population need to take into account different strata in the society (e.g., High income, low income, and medium income), whereas if the curation is about the habit of an individual, it needs to be done over a span of time to consider different situations (say when the person is resting, at work, in stress, and normal control condition) to obtain the unbiased data, which is often difficult due to obvious reasons, including shortage of subjects, human intervention, manipulation, and missing values. Admit all these factors, un-conscious bias could potentially occur as the developer who designed the algorithm or the protocol for study had certain perception to things which at some point of time was misinterpreted by the user or the curator who is making the entry in the dataset could introduce bias in the dataset. Thus fairness of data will always be a question whether it is a large dataset or small, where large datasets might have underrepresentation of certain groups, while small datasets might fail to represent the entire group of data.

10.4.2 Processing Bias

As described in previous section, MBD usually is a huge quantity of unstructured or semi-structured data which could not be analysed using existing database base management system. AI algorithms thus provide software platforms which can reason on inputs to explain the obtained output. Thus processing is divided into pre-processing (acquisition of data), In-processing (AI and ML algorithms), and post-processing (AI and ML based models).

10.4.2.1 Pre-Processing

MBD is the pioneer source of bias, which is introduced to balance the missing values in dataset to balance it before using it to train an algorithm. The notion behind this logic is that "more fair the training data, more reliable the predictive model will be" thus reducing the chances of discrimination and bias in the lineage process. Thus, to achieve this, the data science curators modify the original data by manipulating the class labels for selected observations close to the decision-making factors by using heuristic aiming to carefully balance unprotected and protected groups in training datasets (with loosely controlled effect). Calmon et al. proposed a problematic fairness-aware framework which alters the distribution of data towards fairness, while controlling pre-instance distortion by preventing data utility for learning (Calmon et al. 2017).

10.4.2.2 In-Processing

In-processing approach re-introduces the problem of classification by unambiguously incorporating the discrimination behaviour of model in function via regulation by training on potential target labels. Most of the approaches known so far are true for supervised learning case, which impose equal refurbishment errors for both unprotected and protected groups. Thus selection of right model with appropriate accuracy on large dataset with reduced bias is important to minimise cumulative increase in bias.

10.4.2.3 Post Processing

These approaches focus on the classification model which has been trained using the training dataset thus can be referred to as "learned model". Post processing consist of black box approach (altering the predictions) or white box approach (manipulating the internal parameters of model dataset) (Brault and Saxena 2020). Thus use of AI algorithms with higher level of accuracy (without manipulation of data and inter-relational dependence) might help in dealing with post-processing bias.

However, in recent times, researchers are focusing more on black box approaches rather than white box approaches, which were being supervised by the in-processing methods.

10.4.3 Computing Bias

Computing bias refers to the accountability of an algorithm which is responsible for creation of algorithm, how it functions, and impact of that algorithm on society. During the failure of AI algorithms, the solution is not solved via coding like the trivial times; rather it is rectified and solved using the complex master data and machine learning algorithms. Bias can be computed using bias-aware data collection by explaining the function of AI algorithms and their decisions in simple human terms.

10.4.3.1 Awareness of Bias

Before computing the bias, researchers need to be aware about the pioneer stage of bias; i.e., the data collection stage. There are various models to avoid bias during the data collection, such as mathematical pooling, crowd sourcing, group elicitations, etc. Crowd sourcing relies on significantly large scale collection of data by humans for dealing with missing values in MBD and labelling security in ML algorithms.

Huge sets of data can be collected for a particular scenario (say a normal day in the life of a human) repeatedly over several days and reproducible patterns can be observed (which could be selected as a group of data in a dataset). Sensory data should be checked with manual punch of data thus keeping a check on the dataset and reducing chances of unconscious and technical bias.

10.4.3.2 Modelling Bias

Computing bias demands elucidation and description of meaning, source of collection, notion behind it, model of collection, and the context of bias. Normally, missing data or incomplete categories are considered as bias by the model and replaced by null values, which are considered as negative side effect sources. Thus modelling bias might require deep insight into sources of data, bias, and deep understanding about the working of the algorithms.

10.4.3.3 AI Decisions

Every factor of data annotates something and several factors in a group can lead to an interpretation about that particular situation. Alike AI and ML algorithms interpret the huge MBD datasets to extract knowledge out of them to generate meaningful notions. Generally these decisions are made using specific models and approaches, like black box model, rule based decision sets, model and optimal classification trees, deep neural networks, etc.

Just like every coin has two faces, so are different sides of the outcome. Thus, in upcoming research, we need to develop statistical relational learning to take perspective of knowledge reasoning and accounting, while developing the AI models on more logical grounds.

10.5 Conclusion

We live in a society where primitive research methods have problems of being under-powered, whereas ML, AI, and BDA are over-powered to not only detect the effective size of data that could be of clinical or scientific interest but also meaningful data extract knowledge out of them (Peek et al. 2014).

Data Science (ML, AI, BDA, etc.) has attained a remarkable growth in the last decade. We now have significant knowledge about decision-making algorithms, which could result in decision models based on huge datasets (such as MBD). Big data and Artificial intelligence has tremendous benefits in health and healthcare industry, but noise in medical data might result in false conclusions (Kaplan et al. 2014). With data revolution, we now have an incredible amount of healthcare data stored in cloud storage which is waiting to be analysed. Most of this medical data is unstructured (in the format of graphical, textual, multimedia, etc.) and its original form is of little value (Brault and Saxena 2020). MBD has tremendous variability level of replicability and reproducibility with over-powered analysis leading to false-positive conclusions or biased decision models. Over time, continuous availability of this low quality data with significant noise ratio (error in recording or compromised data quality due to human intervention) might lead to false signals, resulting in wrong inferences. It thus opens a debate for validity of this dataset (Brault and Saxena 2020) for its accuracy before they are used in scientific or clinical research. It is tremendously important (both from ethical and societal points of view) to ask if these algorithms are biased to discriminate on attributes, such as ethnicity, gender, status, etc.

On the orders of former president Mr. Barak Obama, a study was conducted in the United States to explore the role played by data-mining algorithms in decision-making processes. This study concluded that the algorithm and big data analytical technologies tend to cause harm to society way beyond the data privacy. It further added that big data analytical algorithms could potentially display discriminatory results even without discriminatory intent by the developer, resulting in unfavourable situations and disadvantages to needy groups (Williams et al. 2018; Obermeyer et al. 2019; Danks and London 2017).

A major challenge with MBD is its accuracy (data full of biases), limitation in technology (individuals cannot correct their own data) and consistency (lack of standardised protocols). Bias can be acquired from the pioneer source of data (software platform or application or its associated apparatus assisting in data collection; unconscious bias) (Brault and Saxena 2020), during data processing (under- or over-representation of certain group of data which is important for decision making) and post processing (co-relation within the data). Even when these biased attributes are suppressed, algorithm might still discriminate because of inter-dependency within the dataset. Thus, in theory, BDA can eliminate the problems faced by primitive research, however, adding subsidiary challenges considering their overpowered analytical setting.

Sensors embedded mobile technology, such as smartphone, smart devices, and healthcare applications associated with them (mhealth), is gaining popularity in health research. They have made large scale population-based experiments feasible outside of the laboratory setting (Brodie et al. 2018). They have also shared the load of physicians to a certain extent. mHealth is a promising technology to support physicians, just like physicians in clinical setting, which leads to fundamental questions about big data and remote self-reported health outcomes (Recio-Rodríguez et al. 2019; Gorini et al. 2018). To what extent these data are reliable and will mHealth be able to replace more accurate validated clinical examinations data in clinical research? To what extent is the privacy and accuracy of mhealth data maintained from non-validated apps and how appropriate are their outcomes? (Brault and Saxena 2020; Paglialonga et al. 2019).

In a study by Lord et al. (Brodie et al. 2018), they found an extraordinary range of errors in both android and apple devices in comparison to trivial wearable devices, suggesting it as a potential source of unconscious bias occurring from non-validated mobile phone apps (Peek et al. 2014; Wiens et al. 2020). Moreover, there is heterogeneity in the mhealth users (such as different walking speeds, BMI, specific medical condition, etc.) which might lead to systemic bias in MBD, suggesting more efforts in the right direction to come up with platforms which could monitor heterogeneous population around the globe. Thus, when analysing physical big data on a large scale, we should consider unconscious bias against a larger group of individuals. Across globally heterogeneous population, mhealth apps are designed for average consumers (usually considering the factors from the place of origin), thus they are more likely to provide non-validated and biased instrument for monitoring the physiological activities of the body. Thus concluding that greater inaccuracy would be present in a large heterogeneous global population. Despite any discriminating value in the algorithm, uncoil bias may occur due to variability in use of device, heterogeneity in population, etc. (Kaplan et al. 2014; Williams et al. 2018; Obermeyer et al. 2019; Danks and London 2017; Brodie et al. 2018; Wang et al. 2017) This can also be considered as a technical limitation of mhealth technology to provide accurate real-time monitoring and invariability of non-validated health applications to acquire appropriate data to recommend good advice. Big Data have tremendous benefits, but large dataset with noise may cancel out enabling various trends to be observed and this biased big data will eventually lead to false conclusions.

Complexity of predictive algorithms and analytical models may, for instance, limit the capacity to interpret findings of study potentially causing harm when actions are taken upon false predictions (especially on incidental finding). For instance, the information stored in e-Health records is observed data rather than experimental data. Thus, they have a high level of by-systematic bias. Other associated problems for objective nature of Big data, including the fact that interpretations, methods, and inputs are value-driven making it easy to ignore bias, technical quality making unbounded use of data easily justified. Moreover, as the bias is introduced during every stage right from data curation to processing and also algorithm designing and training of algorithm, it cumulatively increases at every stage and adds up to infer completely different outcomes in comparison to actual situation. More research needs to be done on accessing the bias, thus identifying the bias and better predict the outcomes.

References

Arora ASMSM (2018) Advancements in systems medicine using big data analytics. Int J Inf Syst Manag Sci 1(2):13–19

Bibault JE, Giraud P, Burgun A (2016) Big data and machine learning in radiation oncology: state of the art and future prospects. Cancer Lett 382(1):110–117

Brault N, Saxena M (2020) For a critical appraisal of artificial intelligence in healthcare: the problem of bias in mHealth. J Eval Clin Pract. https://doi.org/10.1111/jep.13528

Brodie MA et al (2018) Big data vs accurate data in health research: large-scale physical activity monitoring, smartphones, wearable devices and risk of unconscious bias. Med Hypotheses 119:32–36

Calmon FP, Wei D, Vinzamuri B, Ramamurthy KN, Varshney KR (2017) Optimized pre-processing for discrimination prevention. In: Advances in neural information processing systems 30 (NIPS 2017). Curran Associates, Montreal, pp 3993–4002

Cirillo D, Valencia A (2019) Big data analytics for personalized medicine. Curr Opin Biotechnol 58:161–167. ISSN 0958-1669. https://doi.org/10.1016/j.copbio.2019.03.004

Danks D, London AJ (2017) Algorithmic bias in autonomous systems. Int Jt Conf Artif Intell 17:4691–4697

Denny JC, Van Driest SL, Wei WQ, Roden DM (2018) The influence of big (clinical) data and genomics on precision medicine and drug development. Clin Pharmacol Ther 103(3):409–418

Gorini A, Mazzocco K, Triberti S, Sebri V, Savioni L, Pravettoni G (2018) A P5 approach to m-Health: design suggestions for advanced mobile health technology. Front Psychol 9:1–8

He KY, Ge D, He MM (2017) Big data analytics for genomic medicine. Int J Mol Sci 18(2):1–18

Jensen DM (2018) Harnessing the heart of big data. Physiol Behav 176(1):1570–1573

Kaplan RM, Chambers DA, Glasgow RE (2014) Big data and large sample size: a cautionary note on the potential for bias. Clin Transl Sci 7(4):342–346

Lee CH, Yoon HJ (2017) Medical big data: promise and challenges. Kidney Res Clin Pract 36 (1):3–11

Mitchell TM (1997) Machine learning, 1st edn. McGraw-Hill, New York, NY

Ntoutsi E et al (2020) Bias in data-driven AI systems - an introductory survey. arXiv: 1–19

Obermeyer Z, Powers B, Vogeli C, Mullainathan S (2019) Dissecting racial bias in an algorithm used to manage the health of populations. Science 366(6464):447–453

Paglialonga A, Patel AA, Pinto E, Mugambi D, Keshavjee K (2019) The healthcare system perspective in mHealth. In: m_Health current and future applications. Springer, Cham, pp 127–142

Peek N, Holmes JH, Sun J (2014) Technical challenges for big data in biomedicine and health: data sources, infrastructure, and analytics. Yearb Med Inform 9:42–47

Ramírez MR, Rojas EM, Núñez SOV, de los Angeles Quezada M (2019) Big data and predictive health analysis, vol 145. Springer, Singapore
Recio-Rodríguez JI et al (2019) Combined use of a healthy lifestyle smartphone application and usual primary care counseling to improve arterial stiffness, blood pressure and wave reflections: a randomized controlled trial (EVIDENT II study). Hypertens Res 42(6):852–862
Rumsfeld JS, Joynt KE, Maddox TM (2016) Big data analytics to improve cardiovascular care: promise and challenges. Nat Rev Cardiol 13(6):350–359
Savage N (2012) Digging for drug facts. Commun ACM 55(10):11–13
Saxena M, Saxena A (2020) Evolution of mHealth eco-system: a step towards personalized medicine. Adv Intell Syst Comput 1087:351–370
Saxena M, Deo A, Saxena A (2021) mHealth for mental health. Adv Intell Syst Comput 1165:995–1006
Skilling J, Gull SF (1985) Algorithms and applications. In: Maximum-entropy and Bayesian methods in inverse problems, vol vol. 7. Springer, Dordrecht, pp 83–132
Tang B, Pan Z, Yin K, Khateeb A (2019) Recent advances of deep learning in bioinformatics and computational biology. Front Genet 10:1–10
Wang Y, Sun L, Hou J (2017) Hierarchical medical system based on big data and mobile internet: a new strategic choice in health care. JMIR Med Inform 5(3):e22
Wiens J, Price WN, Sjoding MW (2020) Diagnosing bias in data-driven algorithms for healthcare. Nat Med 26(1):25–26
Williams BA, Brooks CF, Shmargad Y (2018) How algorithms discriminate based on data they lack: challenges, solutions, and policy implications. J Inf Policy 8:78